Daily Warm-Ups GEOMETRY

Common Core State Standards

Jillian Gregory

Thomas Campbell

1 2 3 4 5 6 7 8 9 10
ISBN 978-0-8251-6885-7

J. Weston Walch, Publisher
40 Walch Drive • Portland, ME 04103
www.walch.com
Printed in the United States of America

Table of Contents

Introduction

Daily Warm-Ups: Geometry, Common Core State Standards is organized into five sections, composed of the domains of the Geometry high school conceptual category designated by the Common Core State Standards Initiative. Each warm-up addresses one or more of the standards within these domains.

The Common Core Mathematical Practices standards are another focus of the warm-ups. All the problems require students to "make sense of problems and persevere in solving them," "reason abstractly and quantitatively," and "attend to precision." Many of the warm-ups ask students to develop careful proofs. Students must "use more precise definitions" when proving relationships described. Further, the "correspondence between numerical coordinates and geometric points allows methods from algebra to be applied to geometry." A full description of these standards can be found at www.corestandards.org/the-standards/mathematics/introduction/standards-for-mathematical-practice/.

The warm-ups are organized by standard rather than by level of difficulty. Use your judgment to select appropriate problems for your students.* The problems are not meant to be completed in consecutive order—some are stand-alone, some can launch a topic, some can be used as journal prompts, and some refresh students' skills and concepts. All are meant to enhance and complement high school geometry programs. They do so by providing resources for teachers for those short, 5-to-15-minute interims when class time might otherwise go unused.

*** You may select warm-ups based on particular standards using the Standards Correlations document on the accompanying CD.**

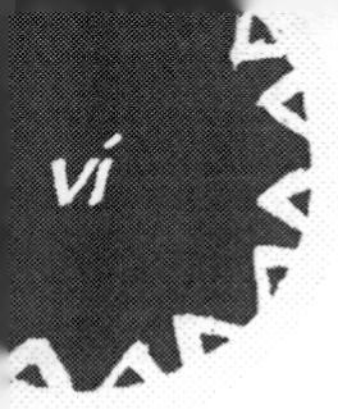

About the CD-ROM

Daily Warm-Ups: Geometry, Common Core State Standards is provided in two convenient formats: an easy-to-use, reproducible book and a ready-to-print PDF on a companion CD-ROM. You can photocopy or print activities as needed, or project them on a large screen via your computer.

The depth and breadth of the collection give you the opportunity to choose specific skills and concepts that correspond to your curriculum and instruction. The activities address the Geometry Common Core State Standards for high school mathematics. Use the table of contents, the title pages, and the standards correlations provided on the CD-ROM to help you select appropriate tasks.

Suggestions for use:

- Choose an activity to project or print out and assign.
- Select a series of activities. Print the selection to create practice packets for learners who need help with specific skills or concepts.

Part 1: Congruence

Overview

Congruence

- Experiment with transformations in the plane.
- Understand congruence in terms of rigid motions.
- Prove geometric theorems.
- Make geometric constructions.

Transformations: Translations

A **translation**, or a slide, is the movement of a figure from one position to another without turning. To the right are examples of a horizontal slide and a vertical slide.

horizontal slide

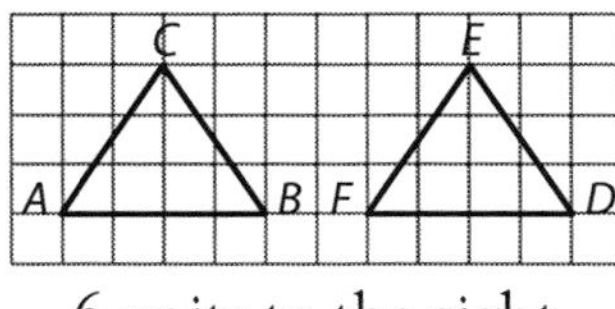

6 units to the right

vertical slide

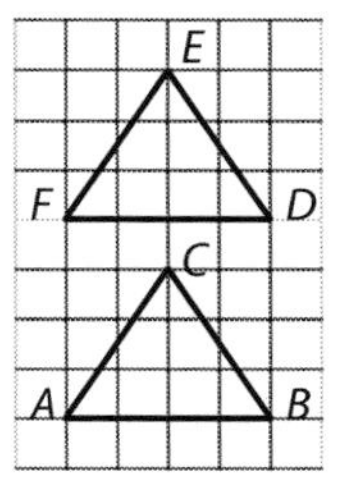

4 units up

Look at the figure below. Slide the figure 4 units to the right and 4 units up. Draw the image on the graph.

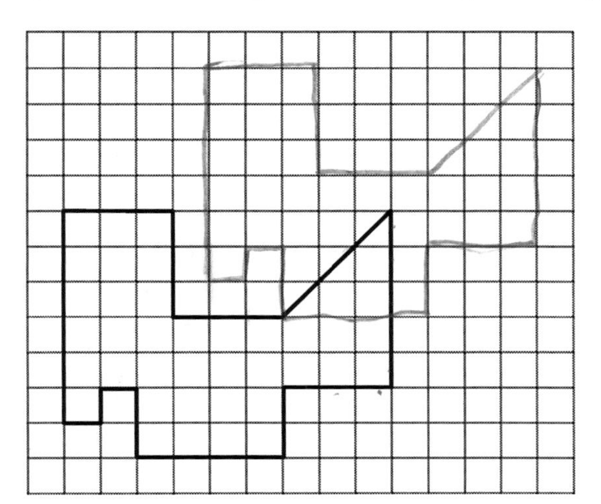

Transformations: Reflections

When a figure is flipped over a line, the mirror image produced is a **reflection.** On the right are two examples of reflections.

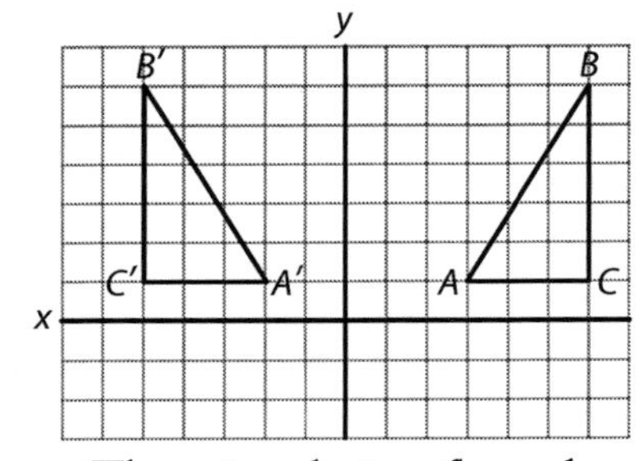

The triangle is reflected across the y-axis.

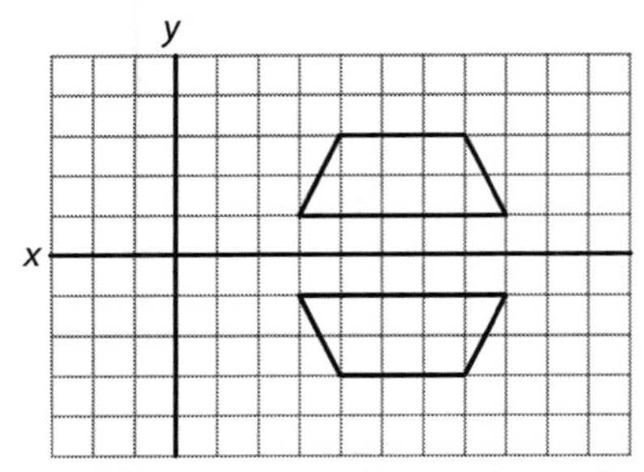

The trapezoid is reflected across the x-axis.

Look at the figure to the right. Reflect the figure across the indicated line of symmetry. Draw the image on the graph.

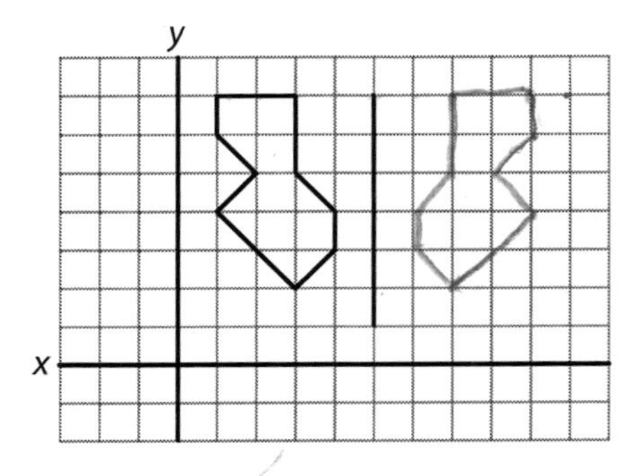

Transformations: Rotations

When you rotate a figure, you turn or spin the figure around a fixed point, the center of rotation.

Example

Rotate the letter *T* 90° at point *P*.

The shaded figure is the original figure.
The unshaded figure is the rotated figure.

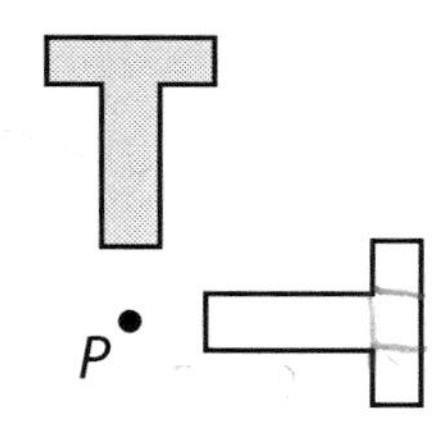

Rotate each figure below about point *P* by the measure indicated.

1. Rotate the figure 45°.

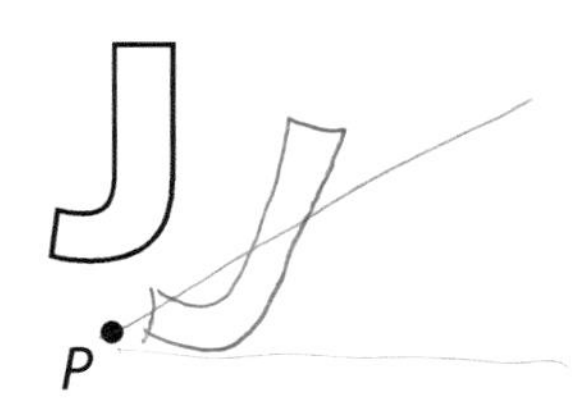

2. Rotate the figure 180°.

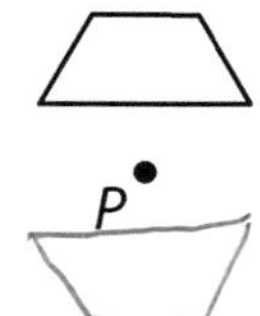

Proof and the Isosceles Triangle

In the diagram below, use what you know about isosceles triangles and congruence to prove that $m\angle 1 = m\angle 2$.

Given: $AB = CB$

BO bisects $\angle ABC$ ($m\angle ABO = m\angle CBO$)

Prove: $m\angle 1 = m\angle 2$

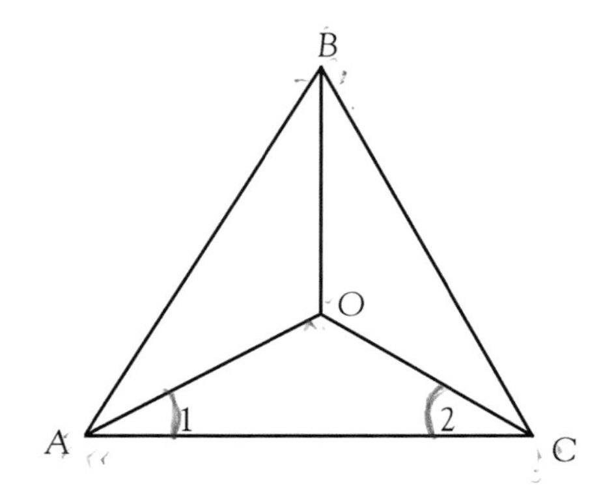

Using SSS Congruence in a Proof

The diagram below is in a shape called a *kite*. A kite is a four-sided shape in which two consecutive sides are equal in length and the other two sides are also equal to each other in length. Prove that the angles formed by unequal sides are equal in measure.

Given: $LO = LR$

$RI = OI$

Prove: $m\angle LOI = m\angle LRI$

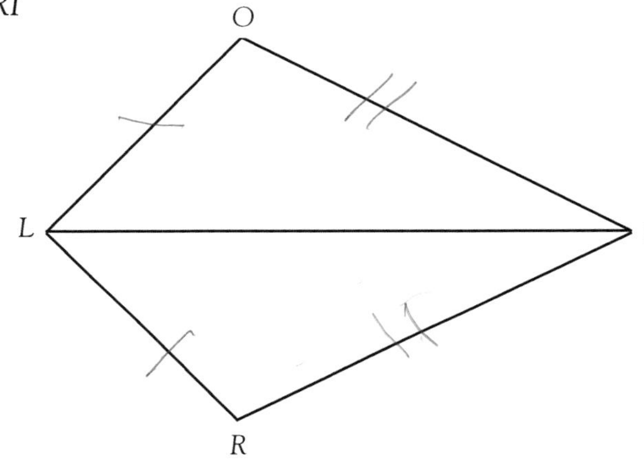

Another SSS Proof

Sometimes the diagrams that support a proof can get complicated. Below is a diagram that has a number of triangles that overlap each other. Use the given information and what you know about congruence and isosceles triangles to prove that $m\angle OMI = m\angle OIM$.

Given: CO = HO

CA = HA

Prove: $m\angle OMI = m\angle OIM$

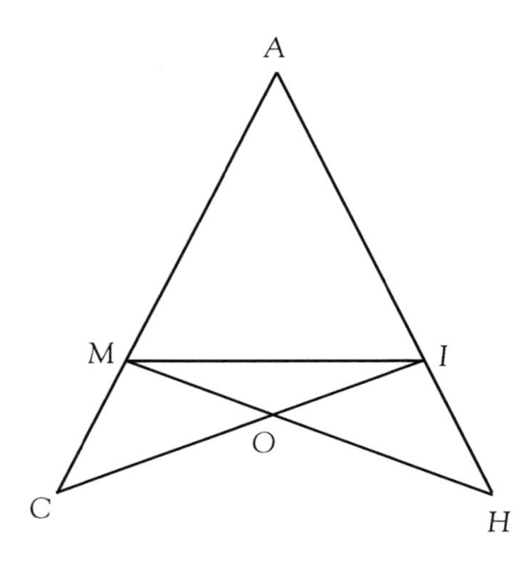

More Proofs with Overlapping Diagrams

Given: $WI = WS$

$IL = SL$ ($WILS$ is a kite)

Prove: $IO = SO$

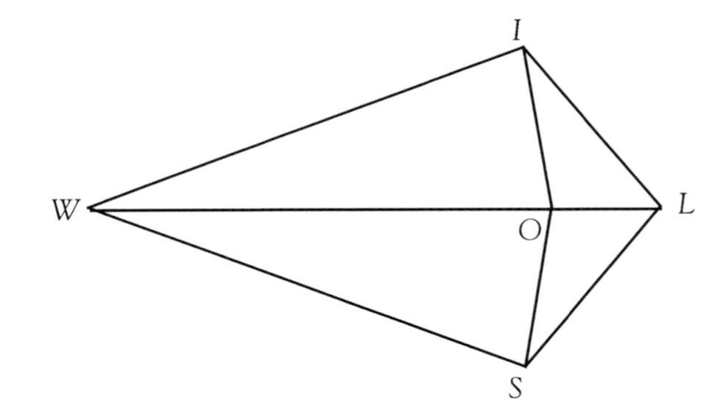

More Proof Practice

Given: $YN = AN$

$OY = OA$

Prove: $RY = RA$

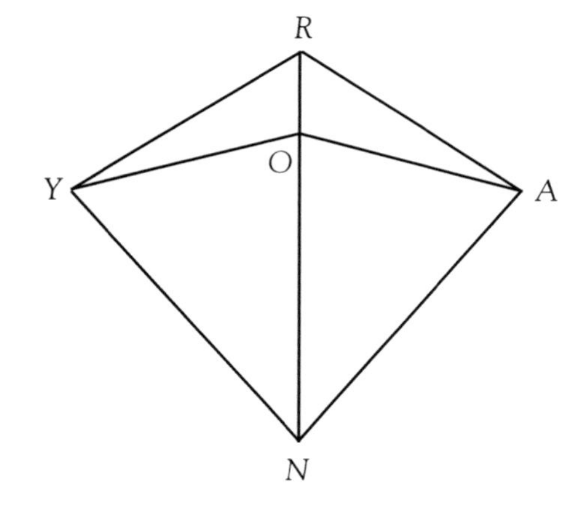

Proving Equal Sides

Given: $SH = SA$

$HN = AN$

$m\angle 1 = m\angle 2$

Prove: $SW = SF$

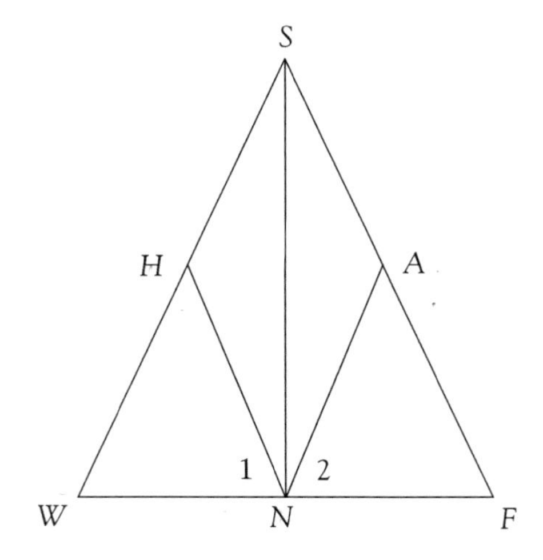

Using AAS

Given: $m\angle IMA = m\angle IAM$

$m\angle MEA = 90°$

$m\angle AGM = 90°$

Prove: $EA = GM$

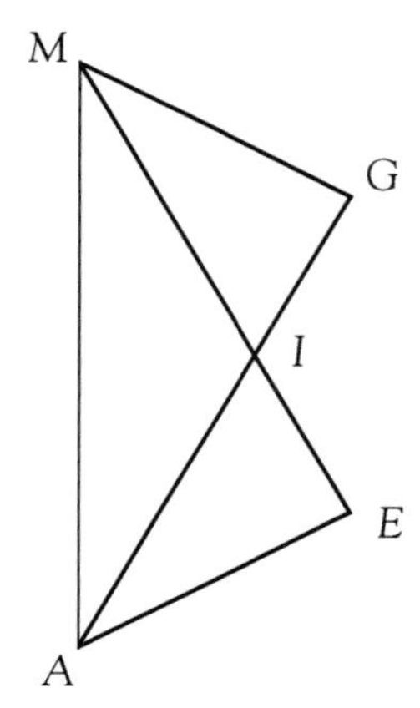

A Proof Using AAS

Given: O is the midpoint of $\overline{EN}$ ($EO = ON$)

$m\angle ILE = 90°$

$m\angle NIL = 90°$

Prove: $EI = LN$

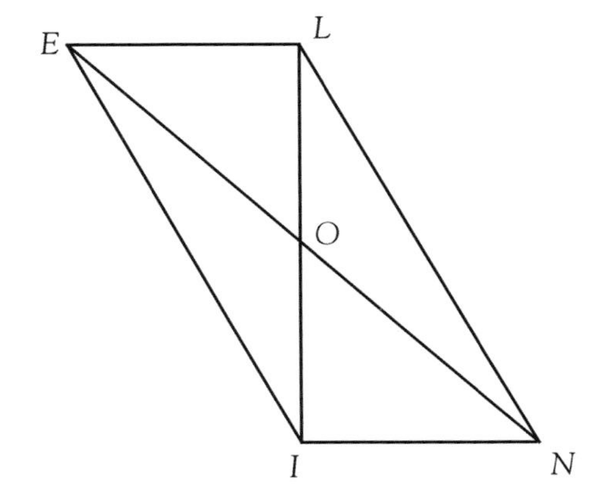

Another Proof Using AAS

Given: $IC = EC$

$m\angle ISE = 90°$

$m\angle ELI = 90°$

Prove: $m\angle OIE = m\angle OEI$

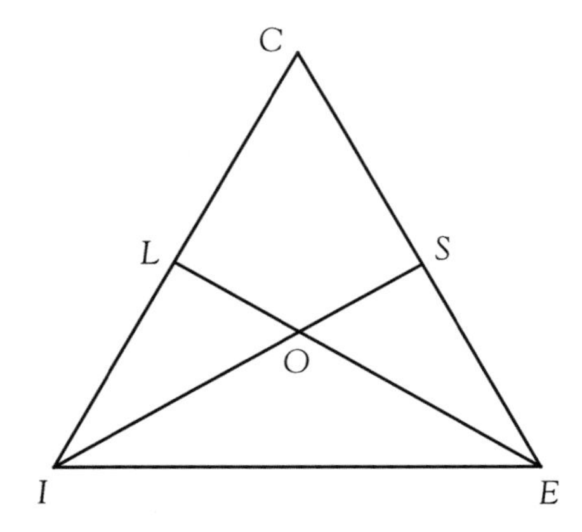

Congruence Through Hypotenuse-Leg Information

Fill in the blanks in the proof below to show that Hypotenuse-Leg information is sufficient to show congruence.

Given: $m\angle BHT = m\angle ENA = 90°$

$BT = EA$

$BH = EN$

Prove: $\Delta BTH \cong \Delta EAN$

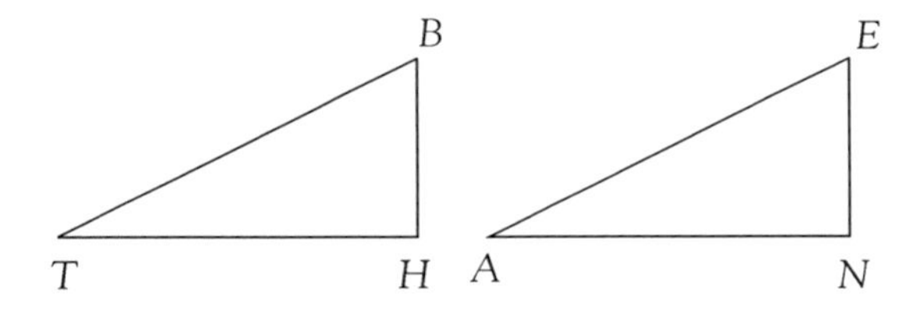

Proof: By copying a line, we may extend $\overline{AN}$ beyond N to Y, such that $NY = HT$. Draw $\overline{YE}$ to create a triangle. We know that $m\angle ENA = 90°$ and that $\overleftrightarrow{AY}$ is straight, so $m\angle ENY =$ ______. Because $HT = NY$, $m\angle ENY = m\angle ENA = 90°$ and $BH = EN$ (given), Δ______ and Δ______ are congruent by ______. Therefore $BT =$ ______. Since BT also equals ______ (given), $EY =$ ______. ΔEAY is now known to be an ____________________ triangle. Using the isosceles theorems, $m\angle EAY =$ ______. From the congruence statement above, $m\angle BTH = m\angle$______, therefore, $m\angle EAN = m\angle BTH$. ΔBTH and Δ______ are now congruent by the AAS congruence theorem.

13

Using the Hypotenuse-Leg Theorem

In the diagram below, use the HL-Congruence theorem.

Given: $AF = GC$

$EF = DG$

$m\angle EAF = m\angle DCG = 90°$

Prove: $m\angle FED = m\angle GDE$

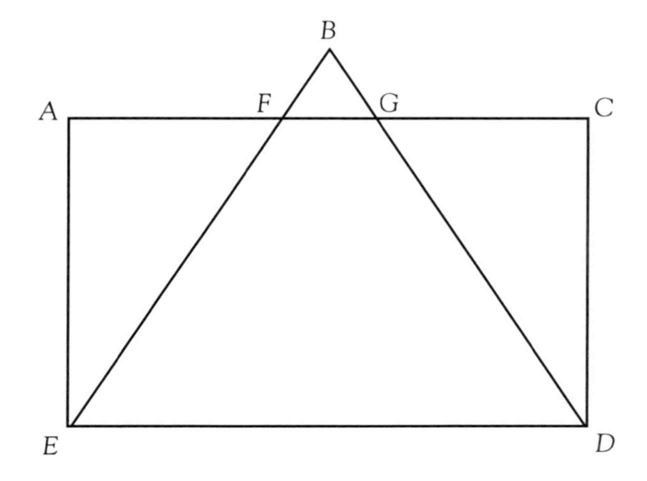

A Proof Using HL Congruence

Given: $AY = RF$

Prove: $LY = LF$

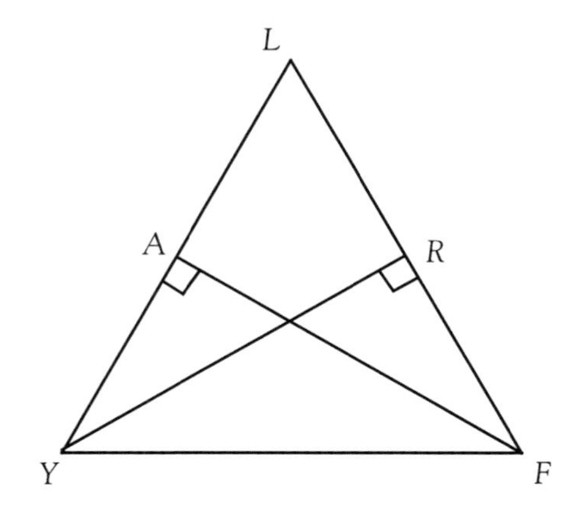

Another HL Proof

Given: $\overline{GO}$ bisects $\angle MGX$

$m\angle WMG = m\angle WXG = 90°$

Prove: $m\angle MOW = 90°$

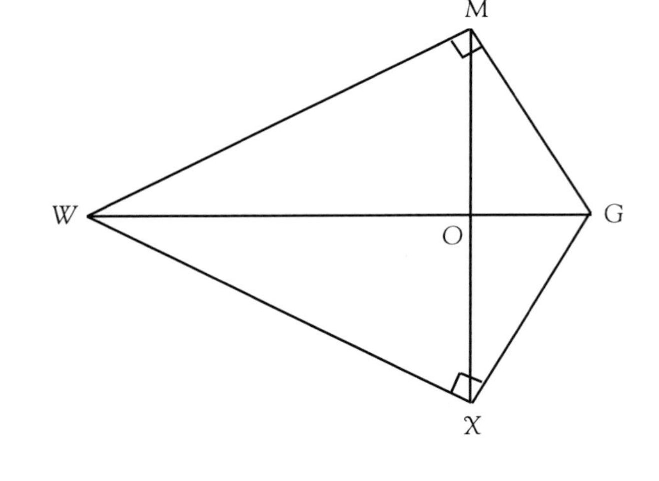

More HL Proof

Given: $SN = SC$

$m\angle SIC = m\angle SKN = 90°$

Prove: $NI = CK$

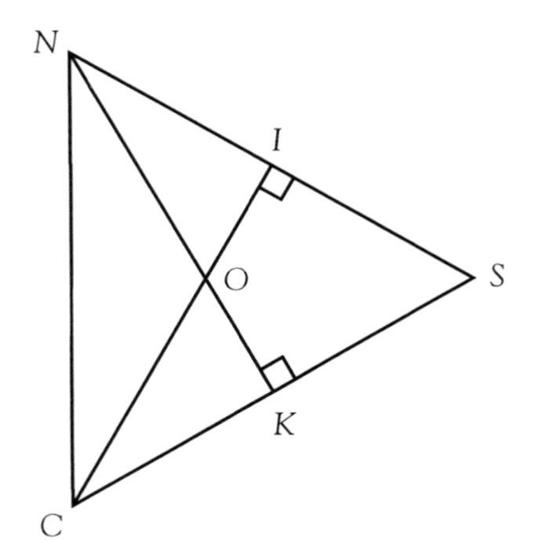

The SSS Method

The Side-Side-Side (SSS) method for proving triangles congruent states: If three sides of one triangle are congruent to the three sides of a second triangle, then the triangles are congruent.

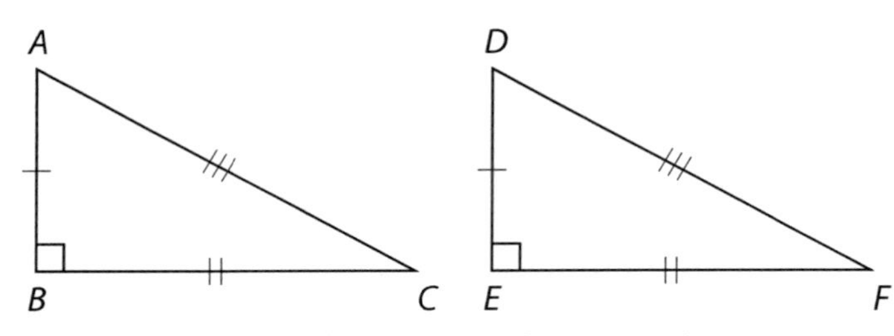

Look at the example above. The triangles are congruent as indicated by the matching number of marks on each side of the triangle.

Indicate whether or not the SSS method can be used to prove that the two triangles given are congruent.

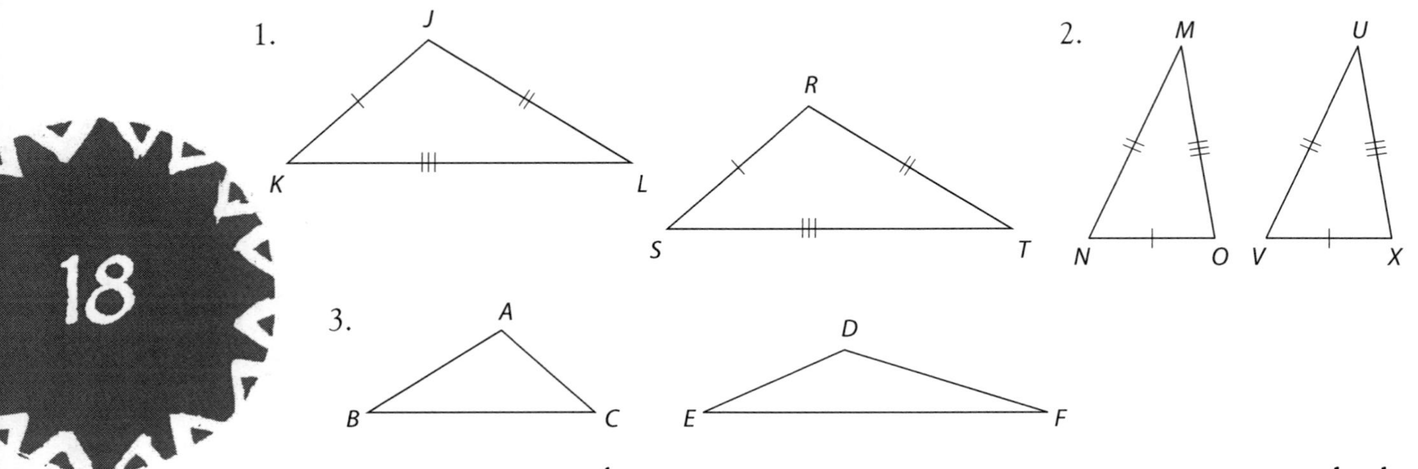

The SAS Method

The Side-Angle-Side (SAS) method for proving triangles congruent states: If two sides and the included angle of one triangle are congruent to two sides and the included angle of a second triangle, then the triangles are congruent.

Indicate whether or not the SAS method can be used to prove that the two triangles given are congruent. Justify your answers.

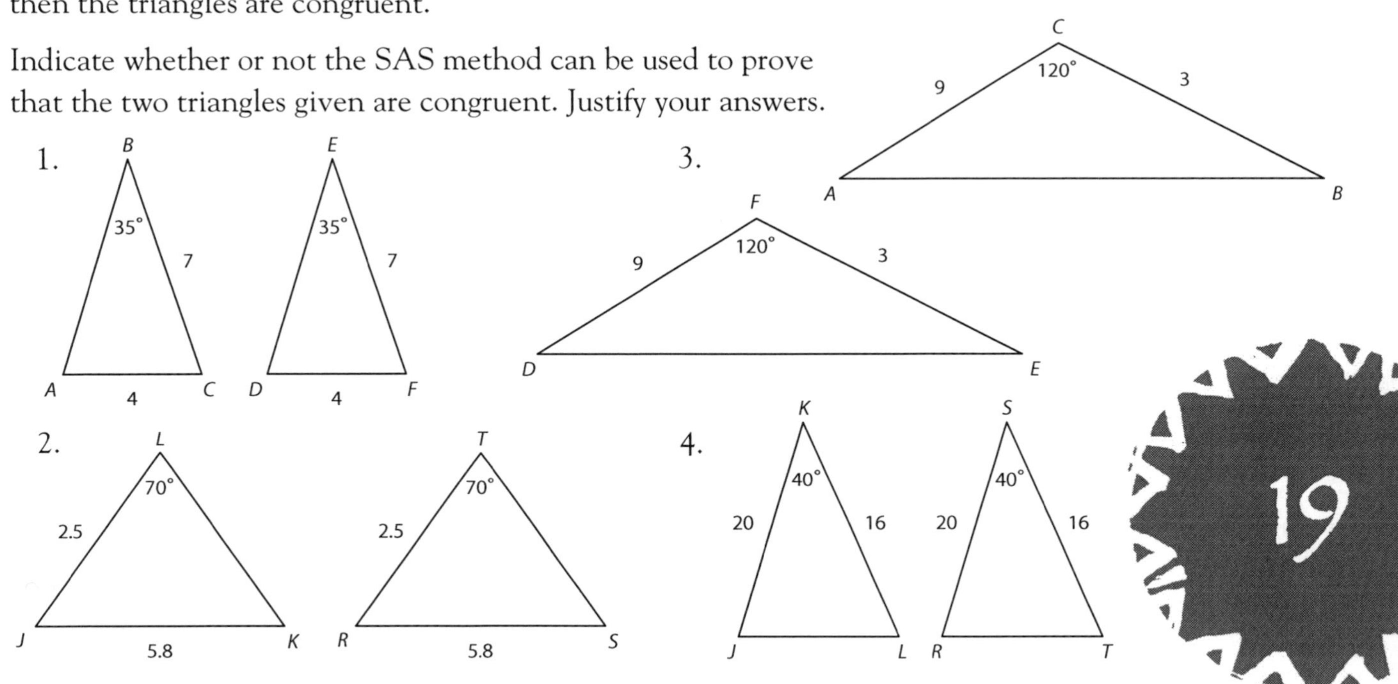

The ASA Method

The Angle-Side-Angle (ASA) method for proving triangles congruent states: If two angles and the included side of one triangle are congruent to two angles and the included side of a second triangle, then the triangles are congruent.

Indicate whether or not the ASA method can be used to prove that the two triangles given are congruent. Justify your answers.

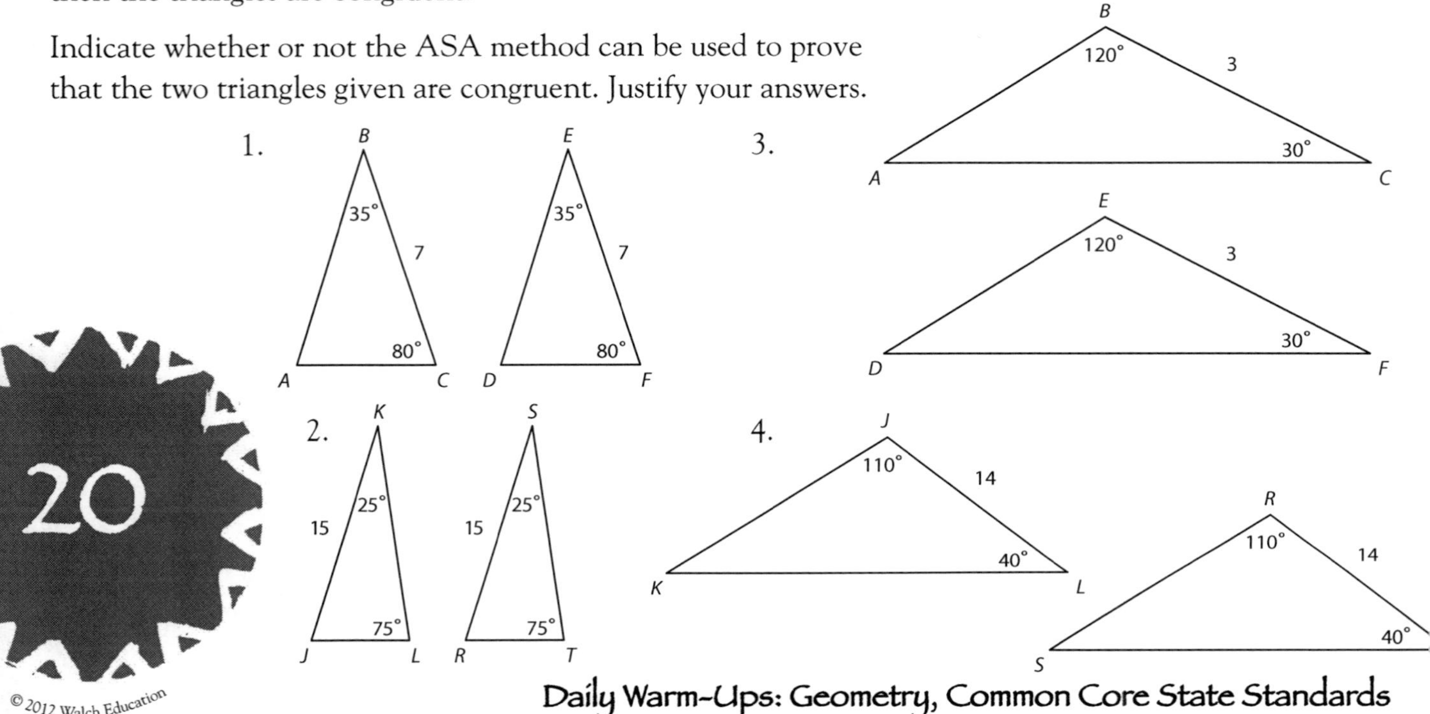

Proving Ground 1

Given: $UE = EC$ (E is the midpoint on $\overline{UC}$)

$m\angle CEL = 90° = m\angle UEL$ ($\overline{EL}$ is perpendicular to $\overline{UC}$)

Prove: $\Delta EUL \cong \Delta ECL$

Remember that we need only show that two sets of sides are equal, respectively, and that the angles formed by those sides are equal in measure.

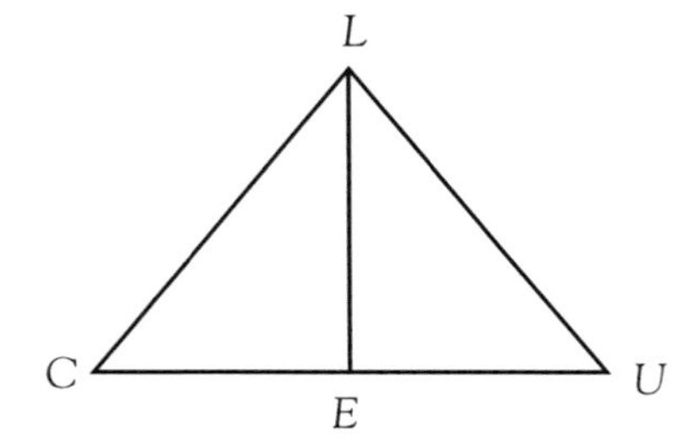

Proving Ground II

Write a paragraph proof. Remember that we can use the SAS theorem.

Given: $GE = EO$

$m\angle GEM = m\angle OEM$

Prove: $GM = OM$

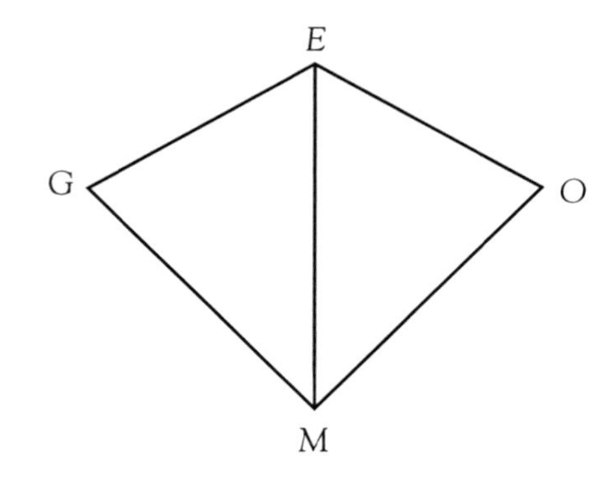

More Proof

In the diagram below, O is the midpoint of $\overline{AB}$ and $\overline{CD}$ ($\overline{AB}$ and $\overline{CD}$ bisect each other). Use a two-column or paragraph proof to prove that AC is equal to BD.

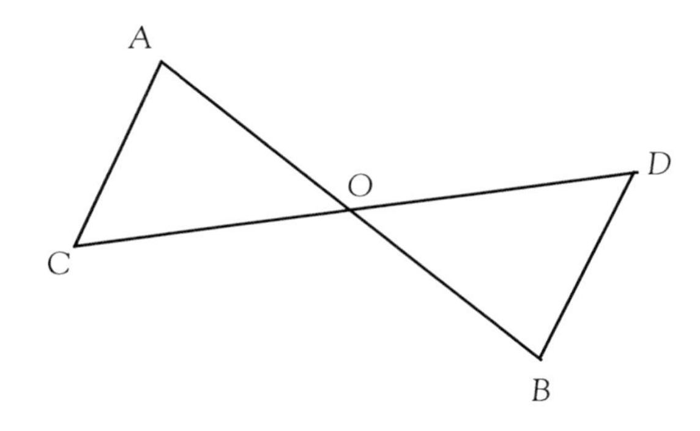

Another Congruence Theorem

Euclid proved that “given two triangles, if you know two sets of angles are equal in measure, respectively, and you know that the included sides are equal in measure, that is enough information to state that the triangles are congruent.” In the five sets of triangles below, which ones do you think are congruent according to the Angle-Side-Angle Theorem?

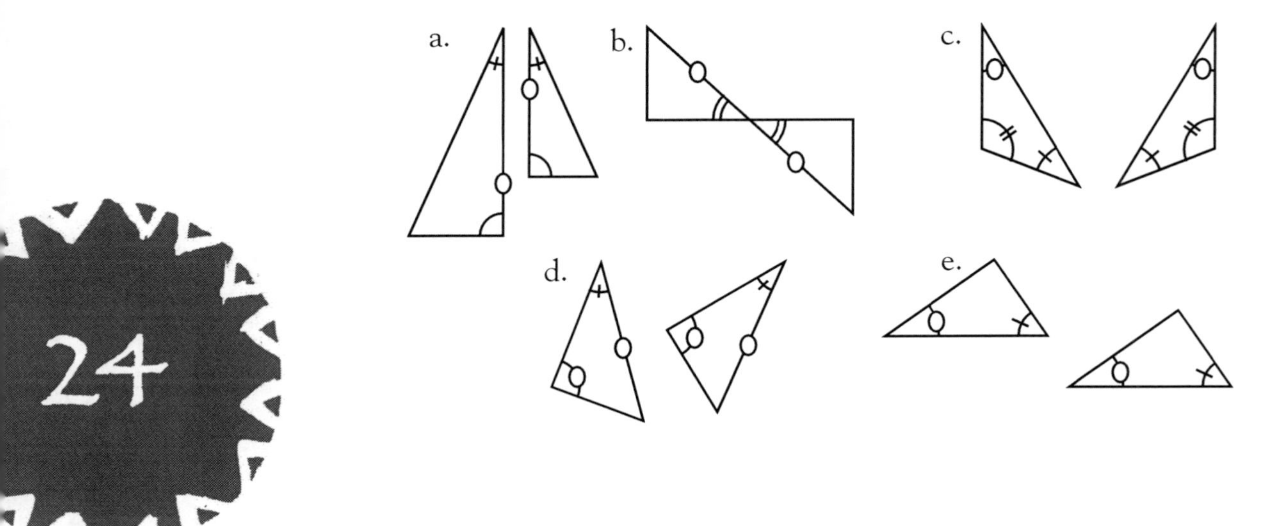

An ASA Proof

Given: $m\angle WAN = m\angle YAN$

$m\angle WNA = 90°$

$m\angle YNA = 90°$

Prove: $WA = YA$

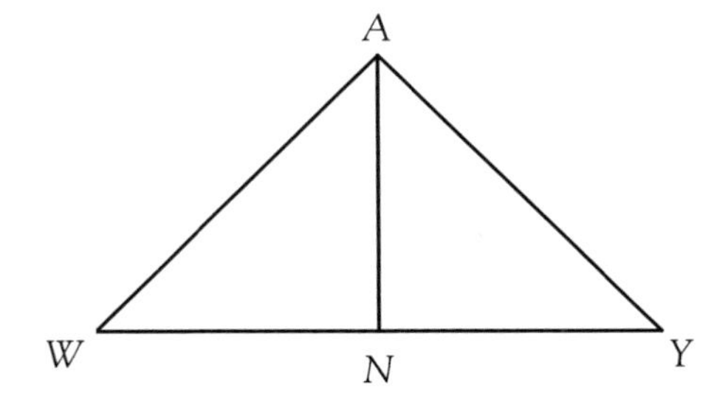

An ASA Proof from Scratch

In the diagram below, $m\angle LOR$ is equal to $m\angle IOR$ (another way to say this is "$\overline{OR}$ bisects $\angle LOI$"). Also, $\angle LRO$ is equal in measure to $\angle IRO$. Use a two-column or paragraph proof to prove that $LO = OI$.

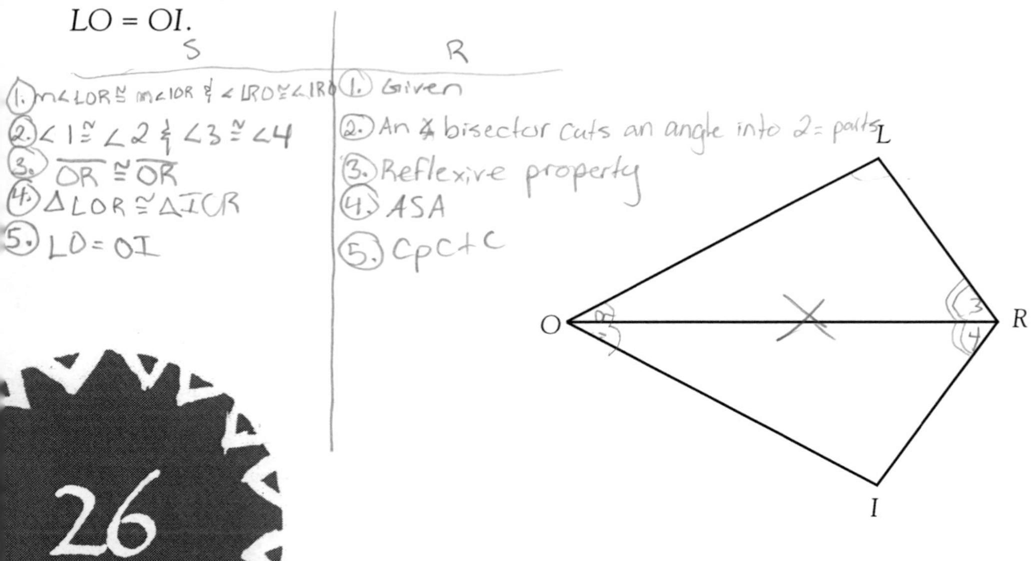

26

Another Congruence Proof from Scratch

Draw and label a diagram based on the information below. State what information is "given" in this problem and what is to be "proved." Using the diagram and the information, write a two-column or paragraph proof of this statement:

"If a line perpendicular to $\overline{XY}$ passes through the midpoint of $\overline{XY}$, and line segments are then drawn from a point on that line to the points X and Y, then two congruent triangles are formed."

Stacking Congruence Statements

Given: $\overline{MA} \perp \overline{EC}$

$GA = AN$

$m\angle EGM = m\angle CNM$

$m\angle EMG = m\angle NMC$

Prove: $ME = MC$

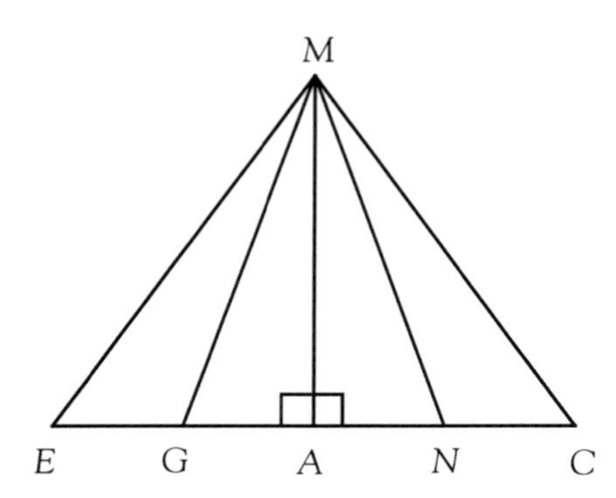

Now You Do One

In the diagram below, you are given that $BH = HY$, $EH = HN$, and $m\angle EBT = m\angle AYN$. Use a two-column or paragraph proof to prove that $BT = AY$.

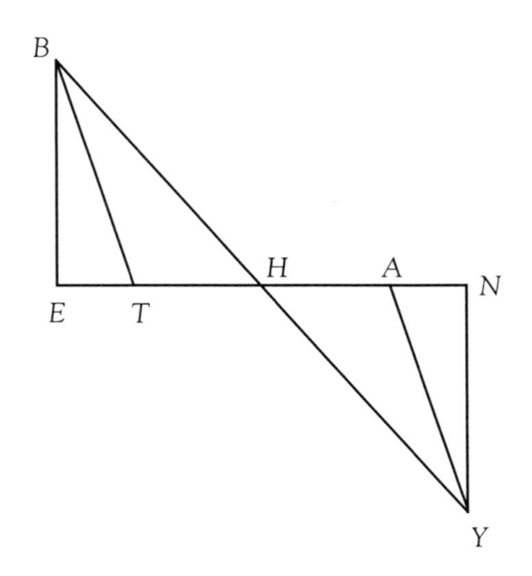

SSS Congruence

Using a ruler, draw a triangle in which you know the length of each side. Report the three side lengths to your neighbor and have him or her draw a triangle with sides of those lengths. Compare your triangles. Experiment with different triangles. Can you create a triangle in which the description of the three side lengths allows your neighbor to draw a triangle different from your own?

The AAS Congruence

We have seen that two triangles are congruent if two sides and the included angle, or two angles and the included side, or all three sides are equal, respectively. By using the sum of the angles of a triangle theorem, we can also show that by knowing two angles and the appropriate *non-included* side, we can prove two triangles congruent. Note the word "appropriate." The known angle at the end of the known side must be the same size angle in each triangle. Prove the AAS theorem by using the diagram below:

Given: $m\angle 1 = m\angle 4$

$m\angle 2 = m\angle 5$

$AB = DE$

Prove: $\Delta ABC \cong \Delta DEF$

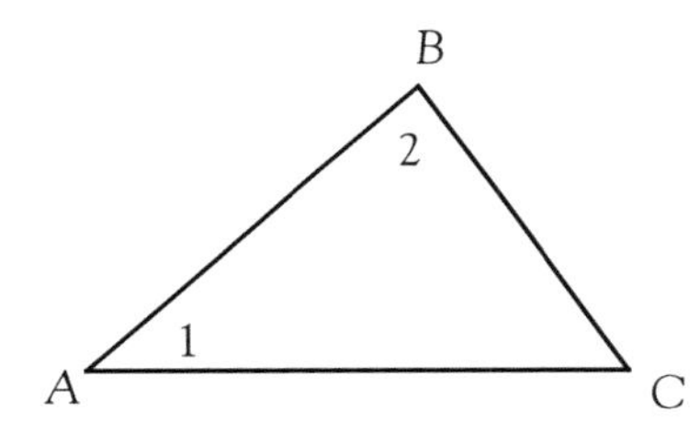

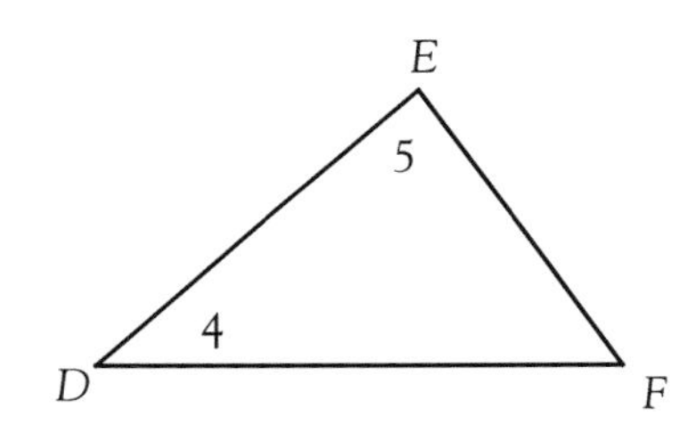

Vertical Angles

Vertical angles are pairs of nonadjacent angles formed when two straight lines intersect.

From the figure on the right, we know the following:

$\angle 1$ and $\angle 3$ are vertical angles and each measures 45°.

$\angle 2$ and $\angle 4$ are vertical angles and each measures 135°.

$\angle 1$ and $\angle 2$ are supplementary angles.

$\angle 3$ and $\angle 4$ are supplementary angles.

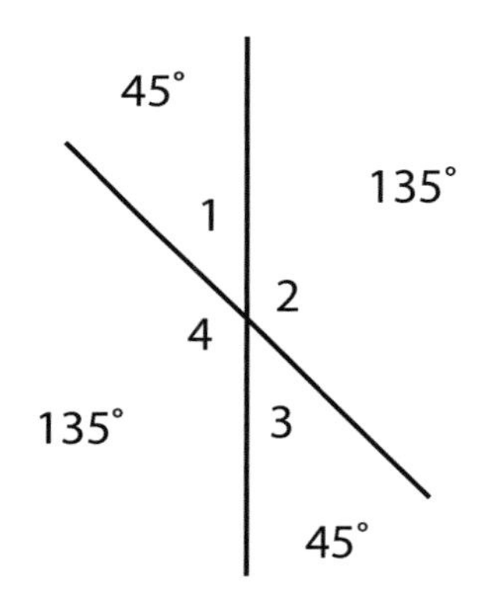

Use the figure on the right to answer the following.

1. Find $m\angle 2$.
2. Find $m\angle 3$.

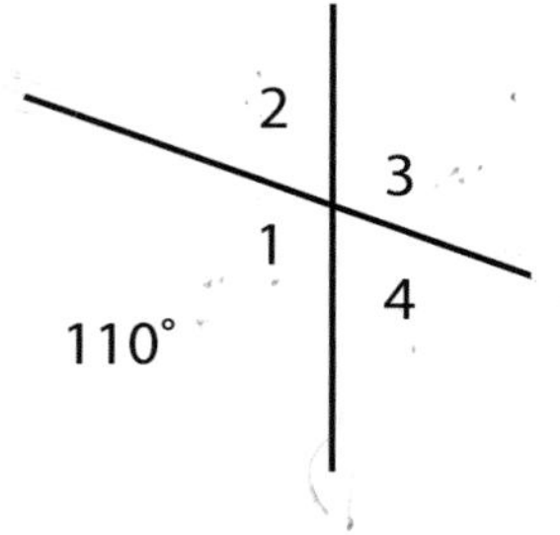

32

Getting Vertical

When two lines meet at a point, the angles formed at that point—by those two lines which do not share a side—are called *vertical angles*. In the diagram below, the angles marked 1 and 3 are vertical angles, as are the angles marked 2 and 4. Remembering that straight angles measure 180°, prove that vertical angles have equal measures in a two-column or paragraph proof.

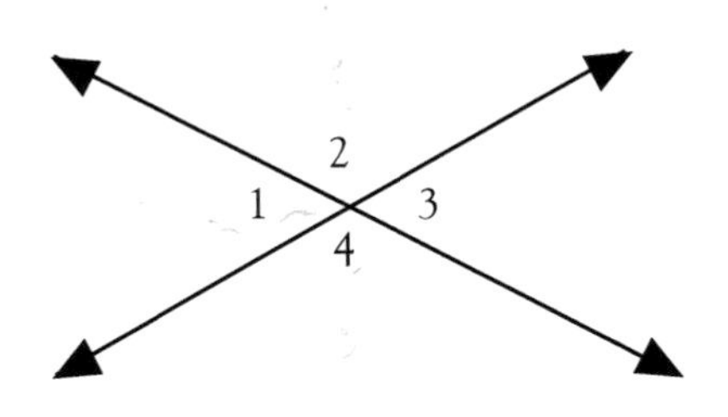

Euclid's Fifth Postulate

Euclid's Fifth Postulate is assumed to be true, but has never been truly proven. The postulate states that, given two lines (l_1 and l_2) and a transversal (t):

- If, and only if, the two angles on the same side of the transversal are formed between the two lines by the transversal and those lines add up to more than 180°, then line 1 and line 2 will meet on the opposite side of the transversal from the angles.
- If, and only if, these two angles sum to less than 180°, then line 1 and line 2 will meet on the same side of the transversal as the angles.
- If, and only if, these two angles sum to exactly 180°, then line 1 and line 2 are parallel.

Using a straight edge and a protractor, draw some pictures and see if you can satisfy yourself that Euclid's Fifth Postulate appears to be true.

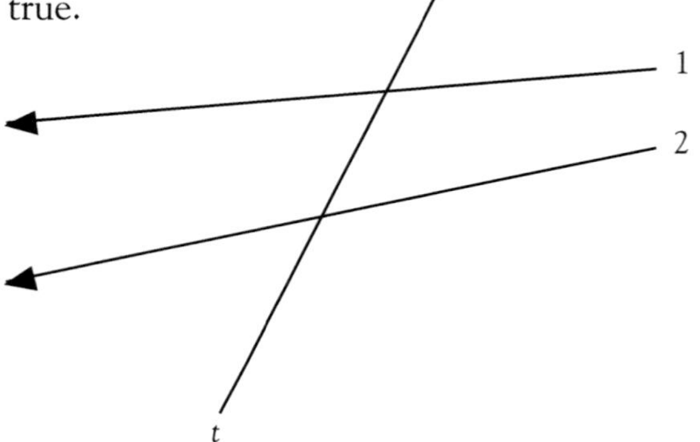

Alternate Interior Angles I

Euclid's Fifth Postulate, or "EV" as it is often called, allows us to do a great deal with parallel lines. Using the diagram below, you will prove a theorem which can be drawn directly from EV. The theorem is: "Given parallel lines and a transversal, the alternate interior angles are equal." *Alternate interior angles* are angles formed by two apparently parallel lines and a transversal.

Given: $\overline{AI} \parallel \overline{EP}$

Prove: $m\angle ATC = m\angle TCP$

$m\angle ECT = m\angle ITC$

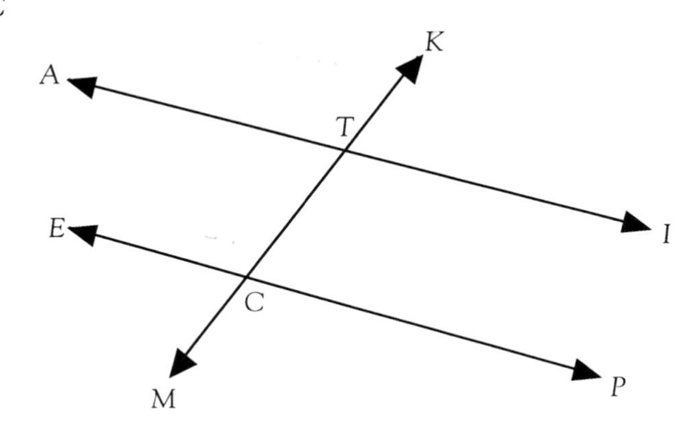

Alternate Interior Angles II

Using the diagram below, see if you can prove the other theorem about alternate interior angles: "Given two lines and a transversal, if the alternate interior angles are equal in measure, then the lines are parallel." This is the converse of the first theorem about these angles. You need only prove this theorem for one of the two sets of alternate interior angles in the diagram.

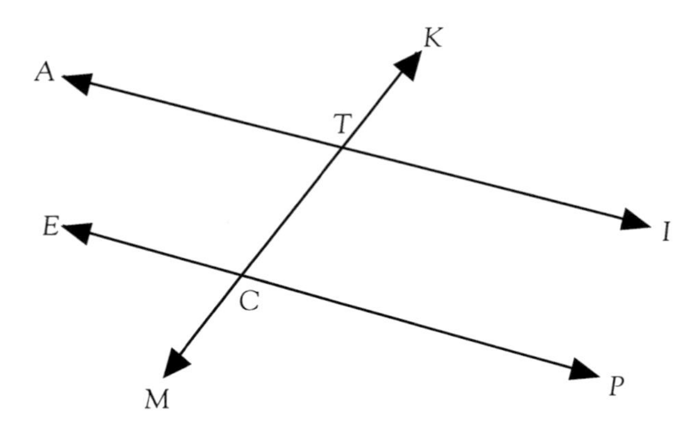

Alternate Exterior Angles I

In the diagram below, $\angle KTA$ and $\angle PCM$ are alternate exterior angles, as are $\angle ECM$ and $\angle KTI$. Use what you know about alternate interior angles and vertical angles to prove, in a two-column or paragraph proof, that if the lines are parallel, then the alternate exterior angles are equal in measure.

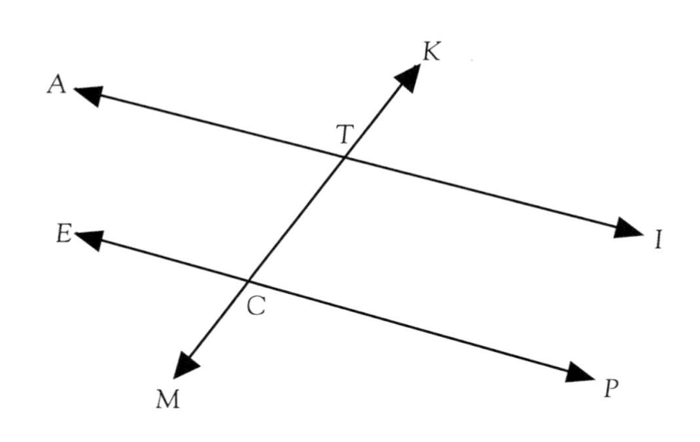

Alternate Exterior Angles II

Prove, in a two-column or paragraph proof, that if the alternate exterior angles are equal in measure, then the lines are parallel. This is the converse of the previous theorem about alternate exterior angles.

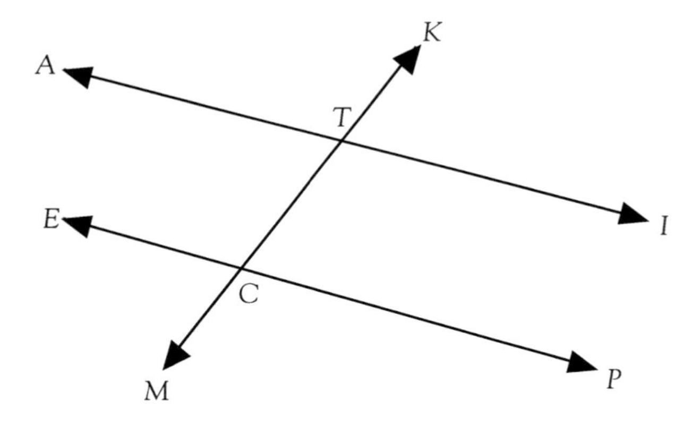

Corresponding Angles I

In the parallel line diagram we have been using, there are other pairs of angles that have special names. $\angle KTA$ and $\angle TCE$ are called *corresponding angles*, as are $\angle ATC$ and $\angle ECM$, $\angle KTI$ and $\angle TCP$, as well as $\angle MCP$ and $\angle CTI$. Using the diagram below, see if you can prove that "given two parallel lines and a transversal, any set of corresponding angles is equal in measure." You need only prove one pair of the angles equal in measure.

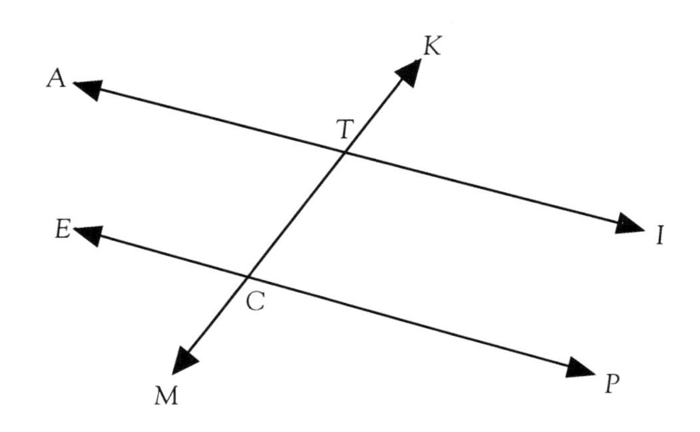

Corresponding Angles II

In a two-column or paragraph proof, prove the converse of "given two lines and a transversal, if any set of corresponding angles is equal in measure, then the lines are parallel." You need prove this for only one pair of the angles.

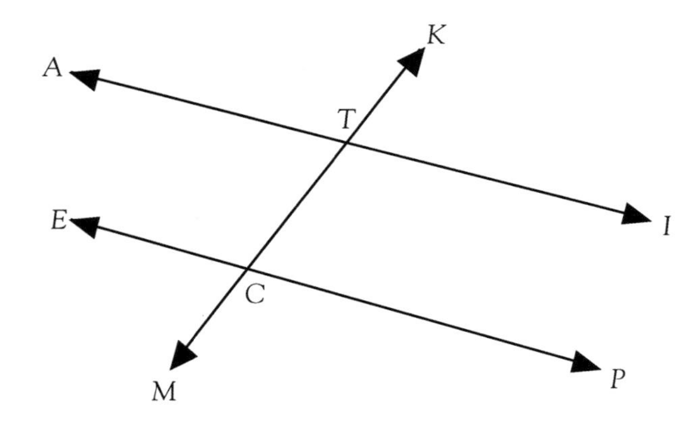

Another Proof

Given: E is the midpoint of $\overline{DF}$.

$\overleftrightarrow{BF}$ is straight.

$\overleftrightarrow{AC}$ is straight.

$\overleftrightarrow{AC} \parallel \overleftrightarrow{DF}$

Prove: $\overline{OA}$ bisects $\angle BOD$.

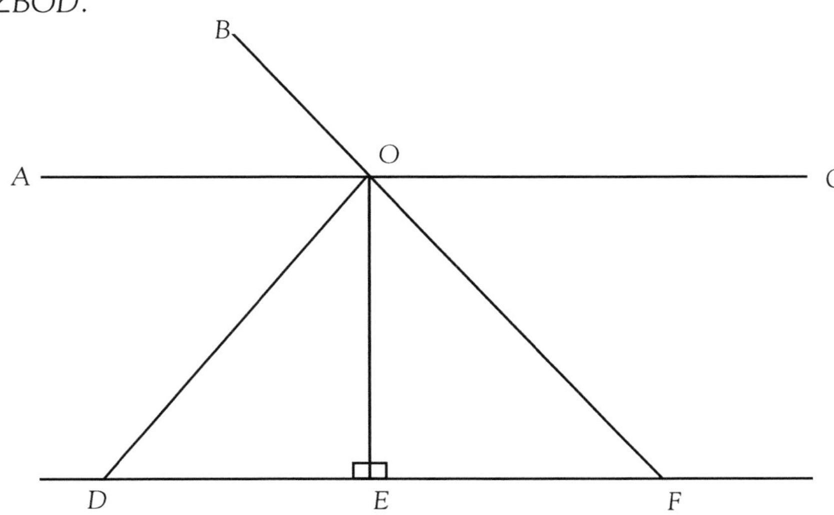

Perfecting Proof Techniques

Given: $\overline{AT}$ bisects $\angle KAE$.

$\overline{AT} \parallel \overline{IE}$

Prove: $AI = AE$

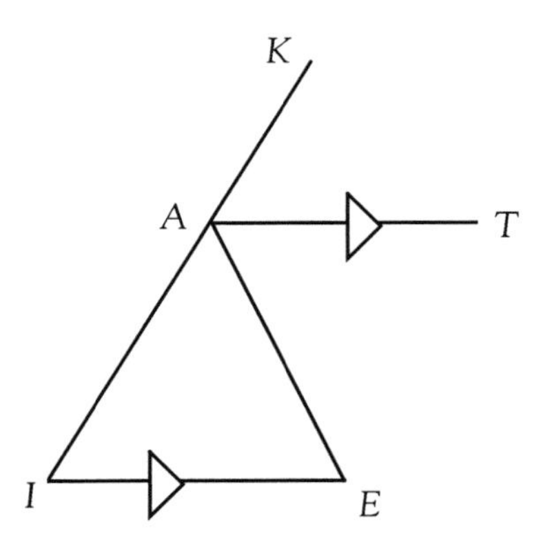

Isosceles Triangles: Median and Altitude

The median of a triangle is a straight line from the vertex of the triangle to the midpoint of the opposite side. The altitude is the perpendicular distance from the vertex of a figure to the side opposite the vertex. The median and altitude of an isosceles triangle are shown below in the two figures.

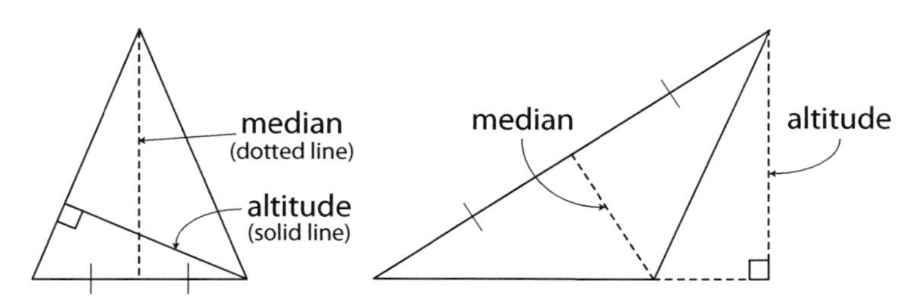

Draw the median and altitude for each isosceles triangle below.

1.

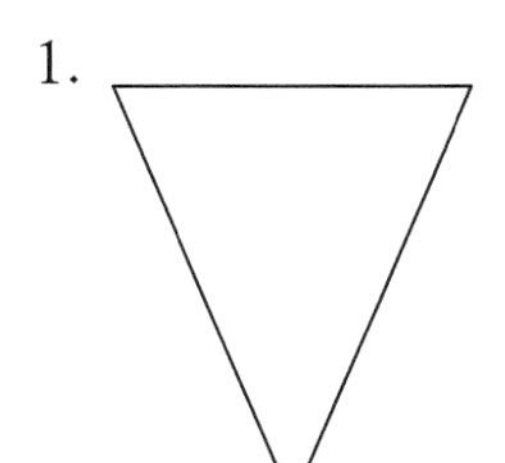

2.

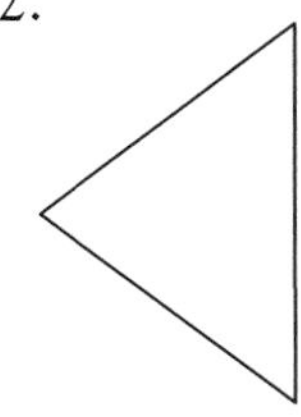

3.

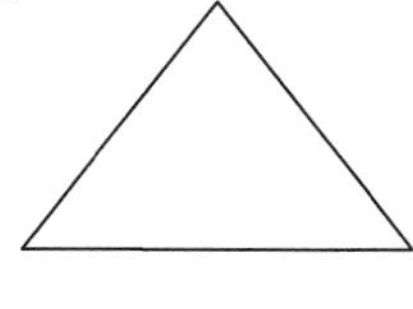

A Balancing Act

Draw a triangle, preferably not a right triangle. Find the midpoints of the three sides of the triangle. Connect each of these midpoints with the vertex of the triangle opposite it. You will find that the three lines, *medians*, are concurrent. Their point of intersection is called the *centroid* of the triangle. The centroid is also the center of gravity of the triangle. If you were to cut the triangle out and rest it on your pencil with the centroid on the pencil point, the triangle would balance. (Be careful of a breeze!)

The Angle Sum Theorem

Draw and cut out a triangle. Number the corners. Now tear off the corners of your triangle. Put the three angles of the triangle down on your desk in such a way that the original vertices of the triangle coincide at a point and the sides of those angles abut one another. What do you notice about the shape that is formed?

Exterior Angle Corollary

A *corollary* to a theorem is another theorem that can be proven using the original theorem. Corollaries usually are not as central an idea and only apply to a specific setting. A corollary to the angle sum theorem has to do with *exterior angles*, the angles formed by one triangle and the extension of one side at one vertex. The corollary states that "an exterior angle of a triangle is equal to the sum of the remote interior angles of that triangle." Another way to say this, in reference to the diagram below, is that $\angle 4$ is equal in measure to $\angle 1 + \angle 2$. See if you can prove this statement true using what you know about the sum of the angles of a triangle and what you know about straight lines.

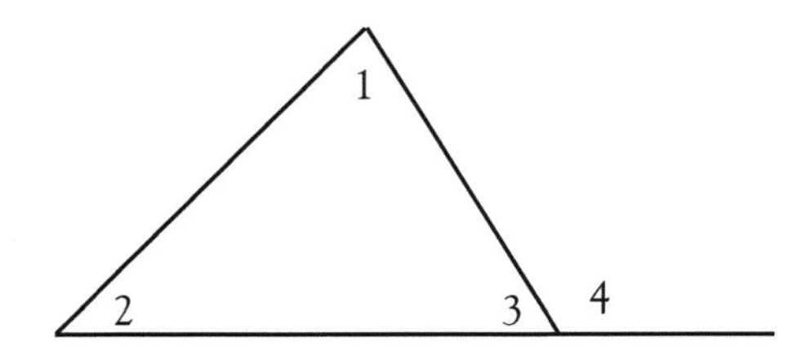

Proving Equal Angles

Use what you know about the sum of the angles of a triangle and vertical angles.

Given: $m\angle 1 = m\angle 2$

Prove: $m\angle 3 = m\angle 4$

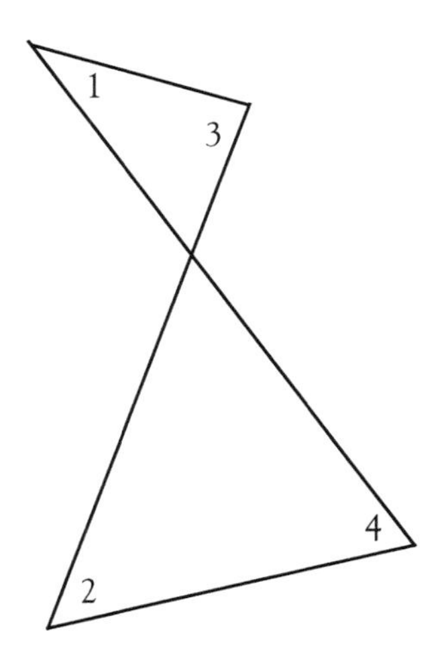

Isosceles Triangles

A triangle with two sides of equal length is called an isosceles triangle. The "opposite angles," the angles which are across from the equal sides, are equal in measure. To prove that two sides being equal also means that two angles are equal, you will need to use SAS congruence and prove the triangle congruent to its own mirror image! To prove that two angles being equal also means that two sides are equal, you can use ASA congruence on the triangle and its mirror image. These are both very short proofs; write them.

Given: $AB = AC$

Prove: $m\angle ABC = m\angle ACB$

Given: $m\angle ABC = m\angle ACB$

Prove: $AB = AC$

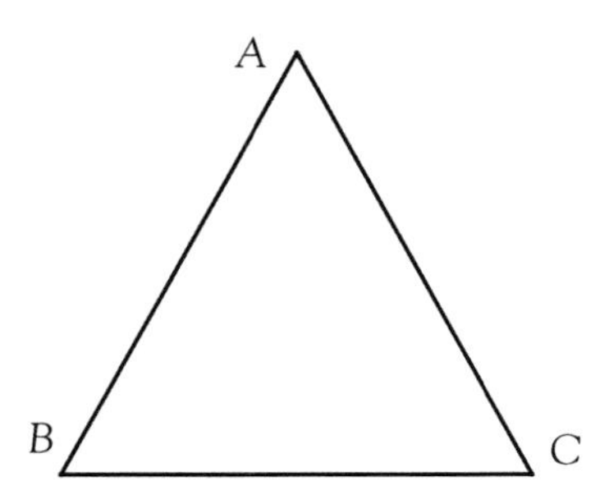

Rectangles

A **rectangle** is a parallelogram that has four right angles. The diagonals of a rectangle are congruent.

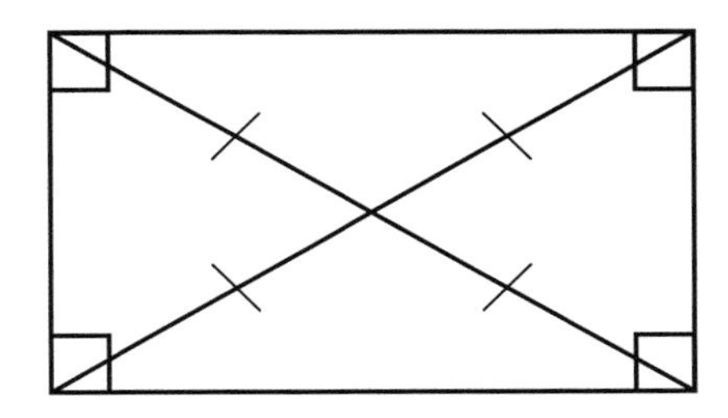

For each statement below, write **T** for true or **F** for false. Justify your answers.

1. All parallelograms are rectangles.

2. All rectangles are parallelograms.

3. All parallelograms have a right angle.

Squares

A **square** is a rectangle that has two congruent adjacent sides. In other words, all sides of a square are congruent.

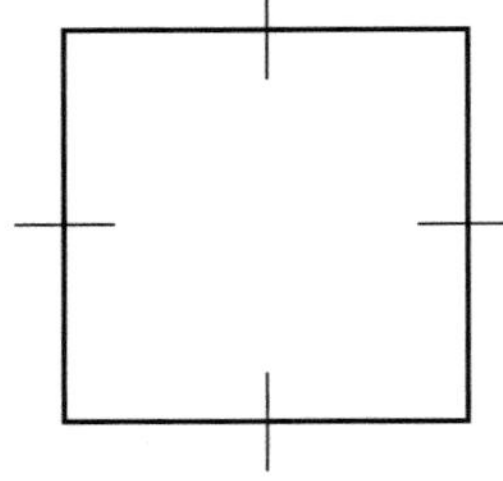

1. For each statement below, write **T** for true or **F** for false. Justify your answers.

   a. All squares are rectangles.

   b. All rectangles are squares.

2. A parallelogram has four right angles and side lengths of x, $x + 2$, $x + 2$, and x. Is the parallelogram a square or a rectangle? Justify your answer.

Rhombuses

A **rhombus** is a parallelogram with two congruent adjacent sides. All sides of a rhombus are congruent, and the diagonals of a rhombus are perpendicular.

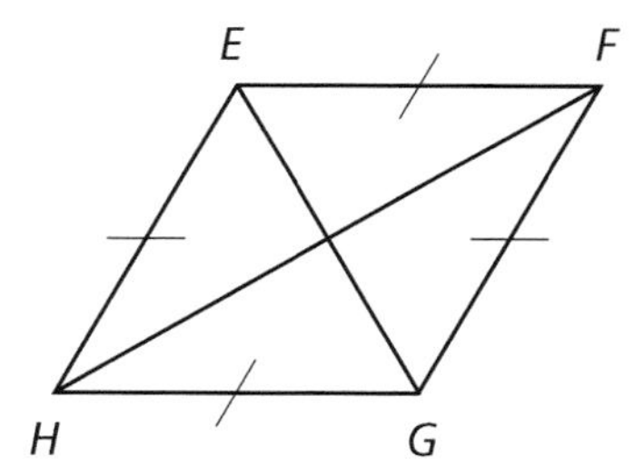

A rhombus may or may not have right angles. For example, a square is a rhombus with right angles.

Use your knowledge of squares, rectangles, and rhombuses to answer the following questions.

1. What type of parallelogram has diagonals that are perpendicular bisectors of each other, but are not congruent?

2. What type of parallelogram has congruent diagonals?

A Proof with Parallel Line Theorems

Write a two-column or paragraph proof.

Given: $m\angle CEF = m\angle BFE$

$\overleftrightarrow{AB} \parallel \overleftrightarrow{CD}$

Prove: $BF = CE$

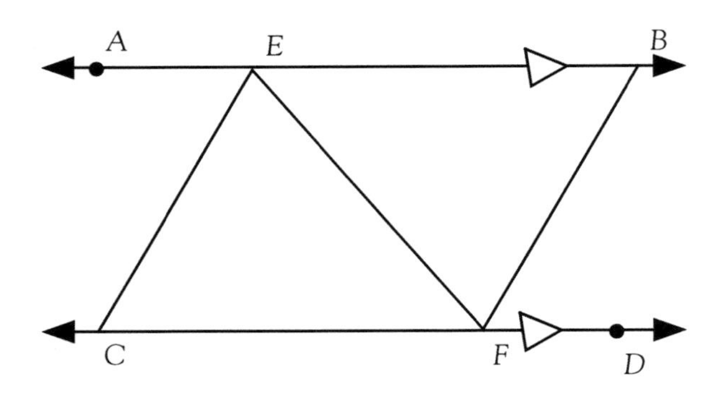

Another Parallel Proof

Given: $\overleftrightarrow{PA} \parallel \overleftrightarrow{LR}$

$PA = LR$

Prove: $\overleftrightarrow{AR} \parallel \overleftrightarrow{PL}$

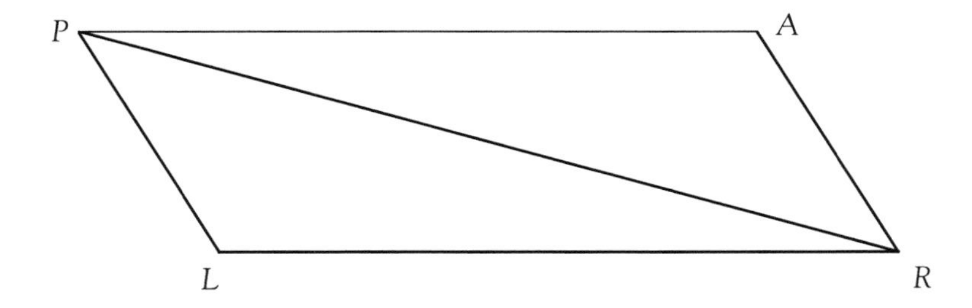

Proving Corollaries

Prove the corollary about parallelograms which states that "given a parallelogram, opposite angles are equal in measure."

Given: $AL \parallel XE$

$AX \parallel LE$

Prove: $m\angle 1 = m\angle 2$

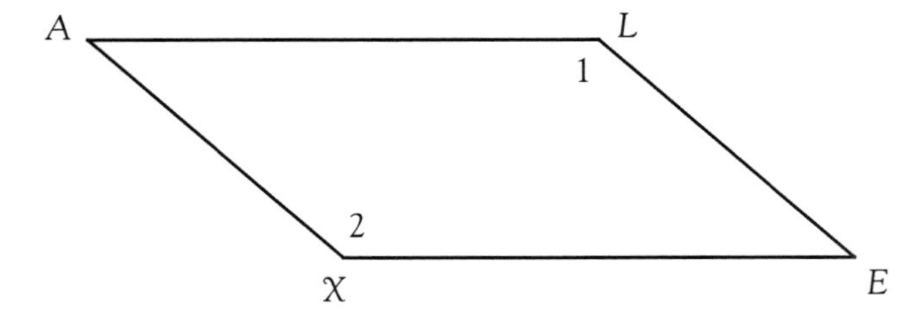

Proving More Corollaries

Given: $FR \parallel AN$

$FA \parallel RN$

Prove: $FR = AN$ or $FA = RN$

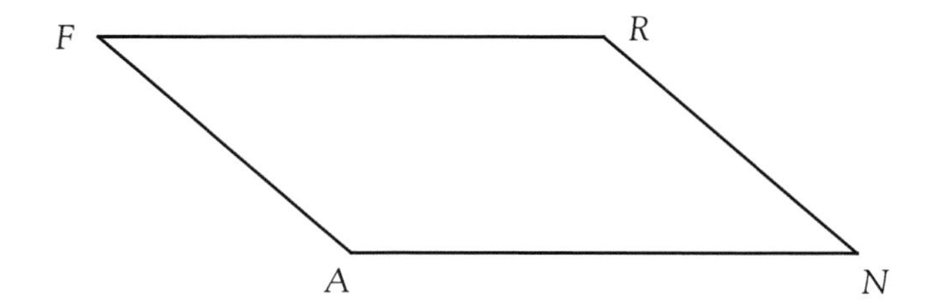

Proving Corollaries with Diagonals

Given: $ZA \parallel CH$

$ZC \parallel AH$

Prove: $ZO = OH$ or $AO = OC$

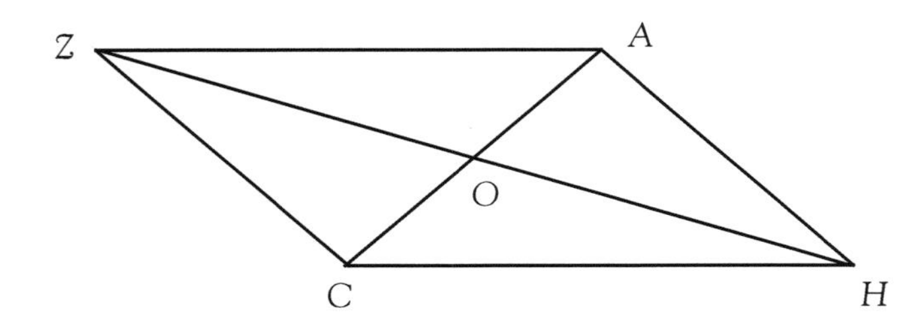

Lines

A line is an infinite set of points. Lines are assumed to be straight. The distance between any two points on a line can be measured. The symbol for a line is $\longleftrightarrow$, which is placed above the name of the line. For example, line CD is written as $\overleftrightarrow{CD}$.

1. Write the name of each line in symbolic form. Then use a ruler to measure the distance between the two points in centimeters.

a. 

b. 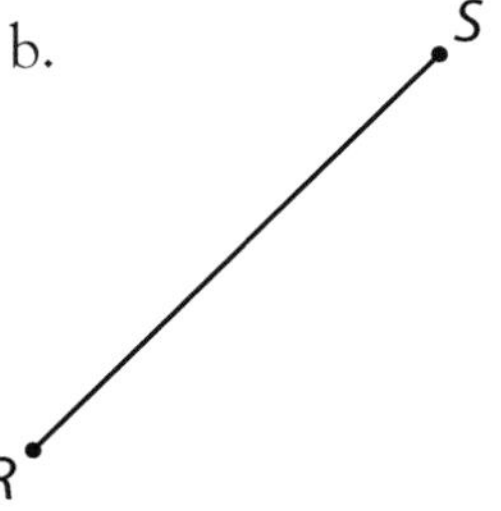

c. Q V

2. Construct each line from the information given.

a. $\overleftrightarrow{GH}$; distance between the points is 12 cm

b. $\overleftrightarrow{AB}$; distance between the points is 5.5 cm

c. $\overleftrightarrow{LO}$; distance between the points is 16.2 cm

Making Copies

Here is a process for copying angles to a new place:

1. Draw a ray to serve as one side of the angle you are copying.
2. Placing the sharp point of your compass at the vertex of the original angle, draw an arc that intersects both sides of that angle.
3. Keeping the same radius on your compass, place the sharp point at the end point of your new ray and draw a semicircle.
4. Going back to the original angle, place the sharp point of your compass on the point where the arc you drew intersects one of the edges of the angle. Open your compass until the pencil point can be placed on the intersection of your arc with the other side of the angle.
5. Going back to your angle copy, maintain the compass radius of step 4. Place the sharp point of your compass on the point where the semicircular arc you drew intercepts your ray. Draw an arc intersecting the previously drawn arc.
6. Using your straightedge, connect the end point of your ray with the point where your two arcs intercept. The angle created is equal in measure to your original angle.

Draw three angles and then copy them.

The Middle of Things

Draw a line segment AB. Use the steps below to bisect $\overline{AB}$.

1. Set your compass to a radius, such that the radius is less than the length of $\overline{AB}$, but more than half that distance.
2. With the sharp point of your compass at A, draw a circle around A using the compass radius from step 1.
3. Maintaining the radius from step 2, put the sharp point on B and draw a circle.
4. The two circles you have drawn intersect at two points (C and D). Use your straightedge to connect C and D.
5. $\overline{CD}$ intersects $\overline{AB}$ at its midpoint M; that is, $AM = MB$.

   (Note that it is also the case that $\overline{CD} \perp \overline{AB}$.)

Making the Cut

Draw an angle. Use the steps below to bisect it.

1. Place the sharp point of your compass at the vertex of the angle and draw an arc that intersects the two sides of the angle.
2. Set the radius of your compass to approximately the distance between where the arc meets the sides.
3. Maintaining this radius, place the sharp point of your compass on one of these intersection points and draw an arc inside your original angle.
4. Maintaining your radius, place the sharp point of your compass on the other point where the arc intersects a side of the angle and draw an arc that intersects the arc you drew in step 3.
5. Connect the vertex point of the original angle to the intersection point created in step 4; this line bisects the original angle.

Drop Me a Line

A line bisector is also perpendicular to the given line. To draw a line perpendicular to any point on your line, follow these steps.

1. Set your compass to a small distance and place the sharp point on the point through which you want to draw your perpendicular. Draw a circle.
2. Find the points at which the circle intersects your original line. Opening the compass a little, put your sharp compass point on one of those intersection points and draw a circle.
3. Keeping the radius of the compass the same, put the sharp point on the other point where your circle met the line and draw an arc that intercepts the arc you drew in step 2.
4. Using your straightedge, draw a line from where these arcs intersect to the original point. This line is the desired perpendicular.

Drop Me Another Line

Follow these steps to draw a line perpendicular to a given line and through a point which is not on that line.

1. Set your compass to a distance greater than the distance from the point to the given line. Place the sharp point on the point through which you want to draw your perpendicular. Draw a circle.
2. Find the points at which the circle intersects your original line. Put your sharp compass point on one of those intersection points and draw a circle.
3. Keeping the radius of the compass the same, put the sharp point on the other point where your circle met the line and draw an arc that intercepts the arc you drew in step 2.
4. Using your straightedge, draw a line from where these arcs intersect to the original point. This is the line through your original point which is perpendicular to the original line.

Equilateral and Equiangular

An **equilateral triangle** has three congruent sides and three congruent angles.

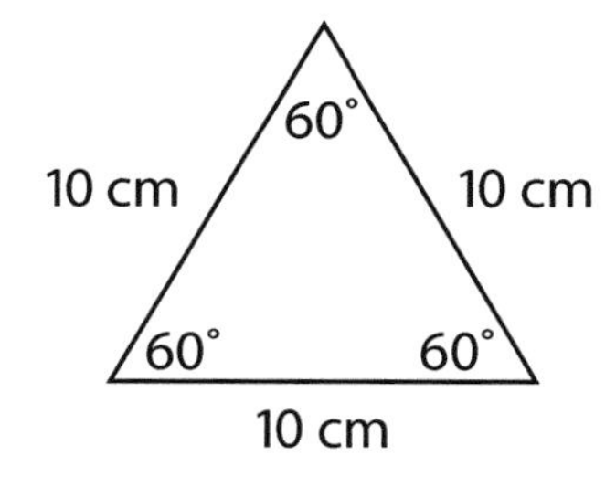

An equilateral triangle is also **equiangular.** This means that if all three sides of a triangle are congruent, then all three angles of the triangle must be congruent, too.

On the other hand, an equiangular triangle is also equilateral. This means that if all three angles of a triangle are congruent, then all three sides of the triangle must be congruent as well.

Use a compass and a straightedge to construct an equilateral triangle inscribed in a circle.

Part 2: Similarity, Right Triangles, and Trigonometry

Overview

Similarity, Right Triangles, and Trigonometry

- Understand similarity in terms of similarity transformations.
- Prove theorems involving similarity.
- Define trigonometric ratios and solve problems involving right triangles.
- Apply trigonometry to general triangles.

Similarity

Two polygons are *similar* if there is some orientation in which each pair of corresponding angles is equal in measure and the ratios of the lengths of each pair of corresponding sides is constant. Intuitively, similarity is like congruence without the shapes being the same size. Below is a picture of a triangle. Using your compass and protractor, draw a similar triangle that is 2.5 times as big. You will need to copy the angles, but make each side 2.5 times as long as it is in the drawing here.

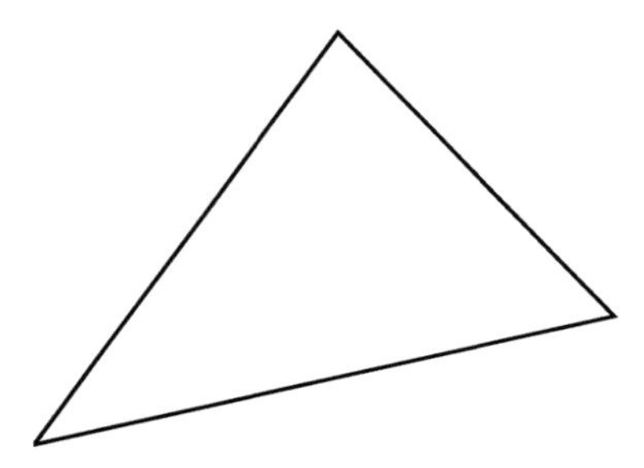

Similar Polygons

Similar polygons have exactly the same shape. Similar polygons must have all pairs of corresponding angles congruent. They must also have all pairs of corresponding sides proportional. Below are some examples of similar polygons.

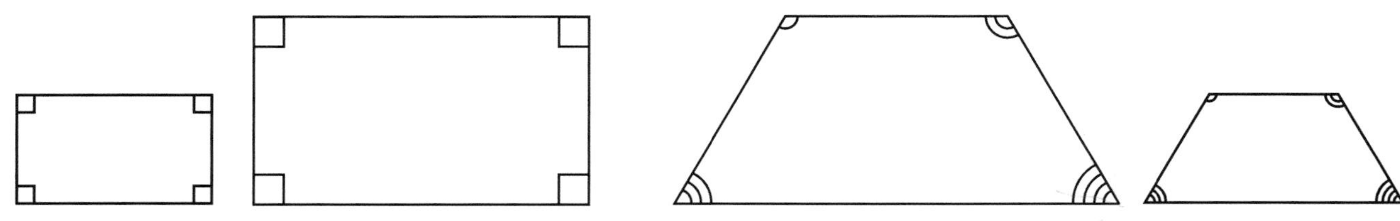

Determine whether or not the two polygons in each pair are similar. Justify your answers.

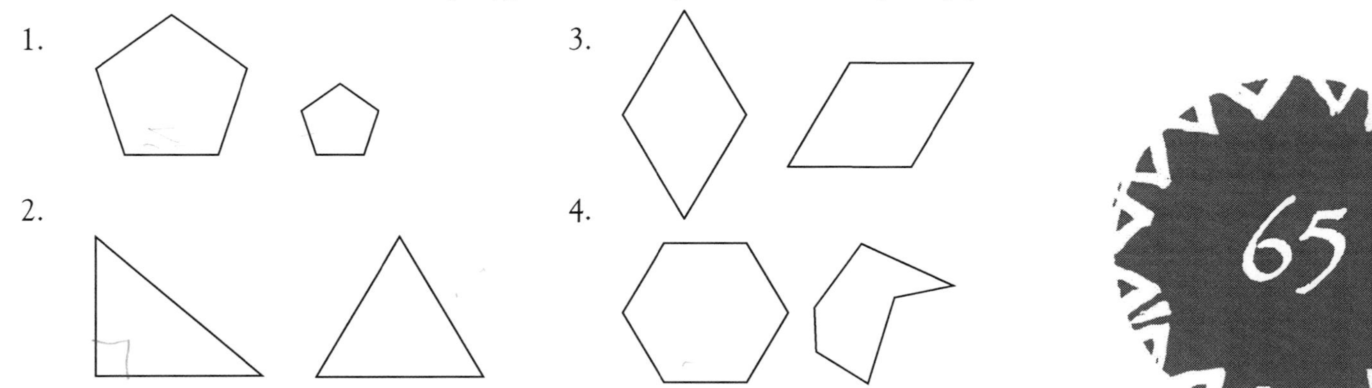

Proving Triangles Similar

The Angle-Angle-Angle (AAA) method states: If the three angles of one triangle are congruent to the three angles of a second triangle, then the triangles are similar. Corresponding sides of similar triangles are proportional. Corresponding angles of similar triangles are congruent.

Use the figure below and your knowledge of similar triangles to answer the following.

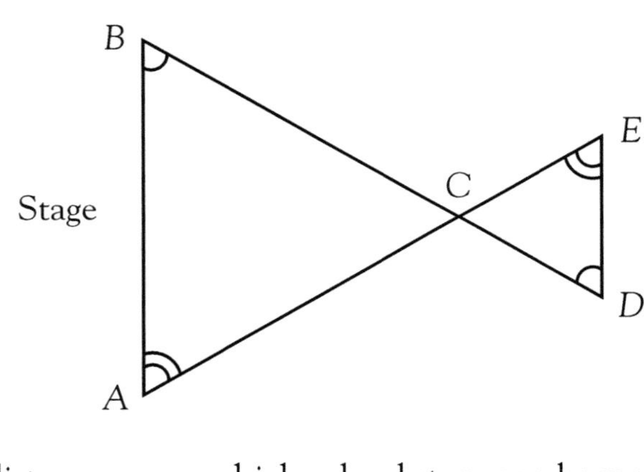

The distance across a high-school stage can be measured using similar triangles. Find AB if $AC = 120$ feet, $DE = 50$ feet, and $CE = 40$ feet.

Using PPP Similarity

Given:

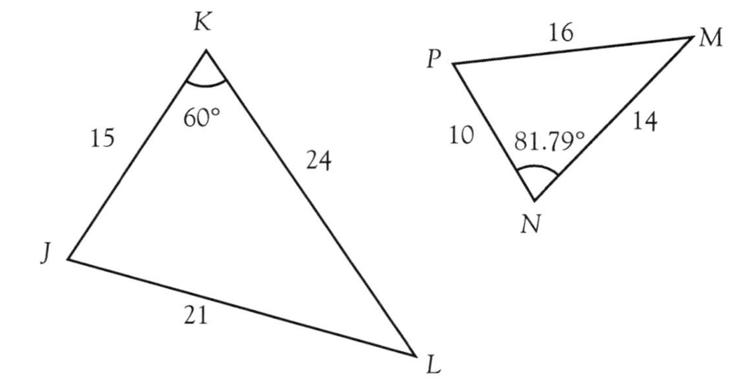

Determine whether or not the triangles are similar. If so, find the measure of $\angle NMP$.

AA Similarity

In order for two triangles to be similar, they must have some orientation in which all three pairs of corresponding angles are equal and all three pairs of corresponding sides are in constant ratio. However, as with congruence, some of that information is enough to conclude similarity. In fact, knowing that two pairs of angles are equal in measure allows you to conclude that the triangles are similar. Use what you know about ratios to determine the unknown lengths *AN* and *EM*.

Given: $m\angle AEG = m\angle MNA = 90°$

$EG = 4$, $MN = 7$, $AG = 5$, $EA = 3$

Find: *AN* and *EM*

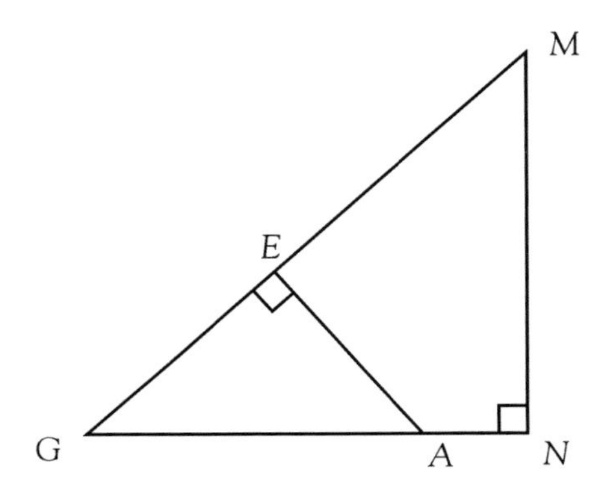

Parallels and AA Similarity

In the diagram below, the bases of the two overlapping triangles are parallel. Because of our theorems about parallel lines, you can prove that the triangles are similar.

Given: $\overline{BC} \parallel \overline{DE}$

Prove: $\Delta ABC \sim \Delta ADE$

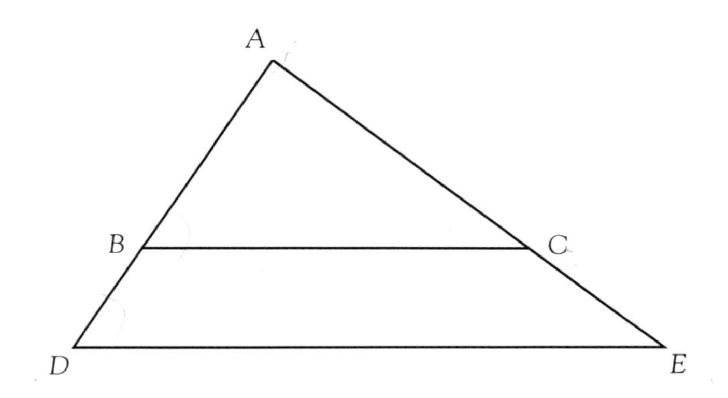

Using AA to Find Lengths

Using what you know about parallel lines and the AA similarity theorem, find BD.

Given: $AB = 3\sqrt{3}$

$AC = 4\sqrt{2}$

Find: BD

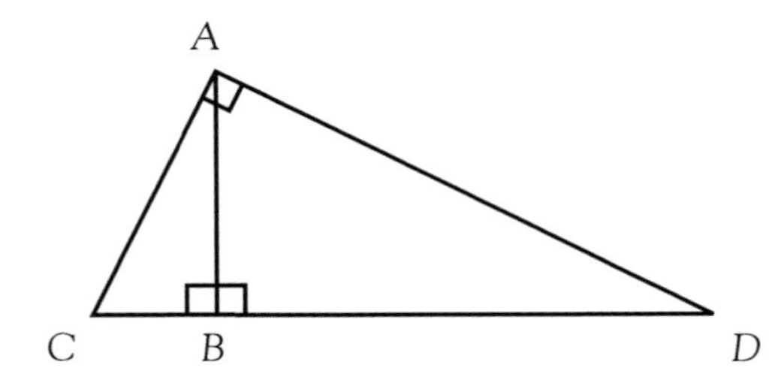

Proof with AA Similarity

Given: $\overline{AB} \parallel \overline{DE}$

Prove: $\Delta ABC \sim \Delta EDC$

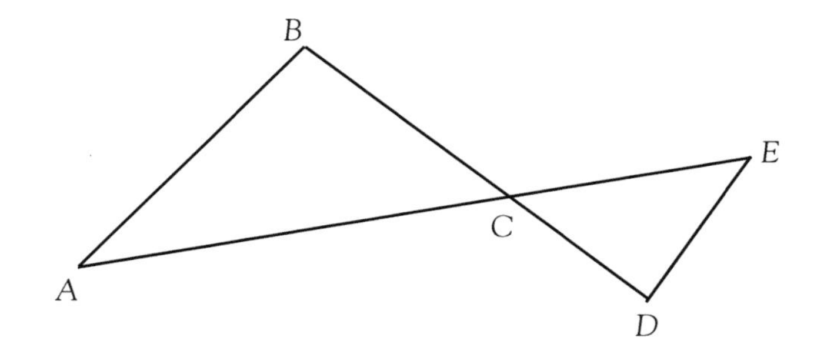

More Proof with AA Similarity

Given: $\overline{AC} \perp \overline{BD}$

$m\angle 1 = m\angle 2$

Prove: $\Delta BAC \sim \Delta EDC$

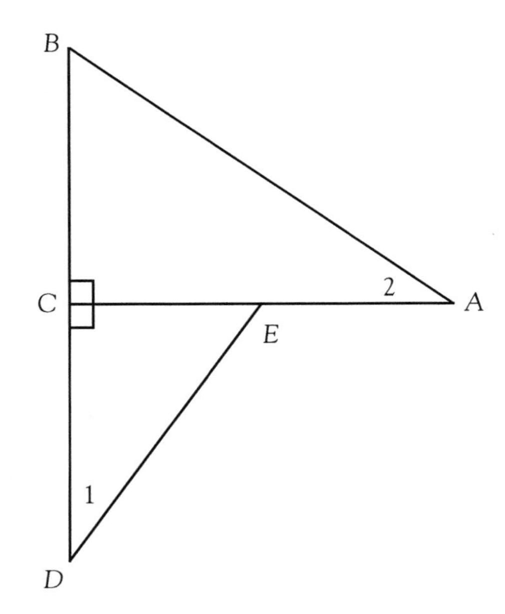

Parallelograms and AA Similarity

Given: *ABCD* is a parallelogram.

Prove: $\Delta BDA \sim \Delta DBC$

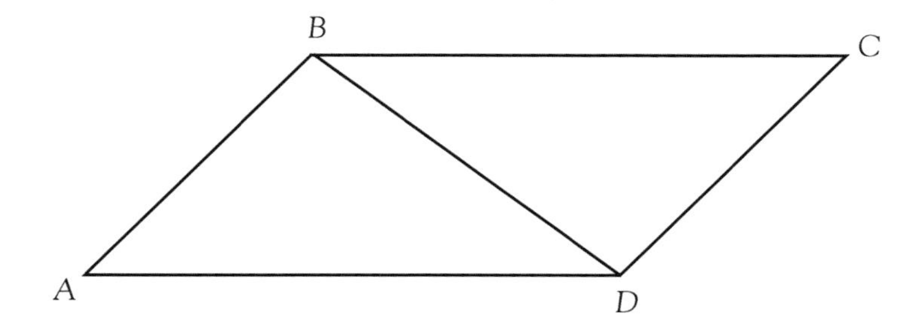

Trapezoids and AA Similarity

A trapezoid is a four-sided figure in which only one pair of opposite sides are parallel. Of course, the parallel sides help get you equal angles, which in turn can be used for AA similarity. Use this strategy to prove the following.

Given: $\overline{RS} \parallel \overline{VT}$

Prove: $\Delta ROS \sim \Delta TOV$

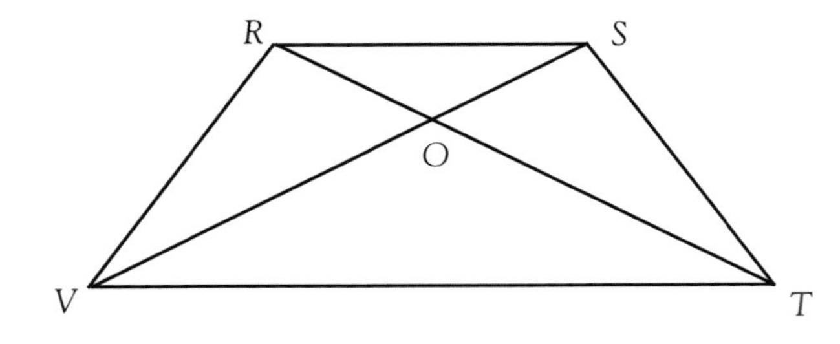

More AA Proof

Given: *ABCD* is a parallelogram.

Prove: $\frac{AB}{ED} = \frac{BF}{DA}$

(Remember: If the triangles are similar, then the ratio of their sides is constant.)

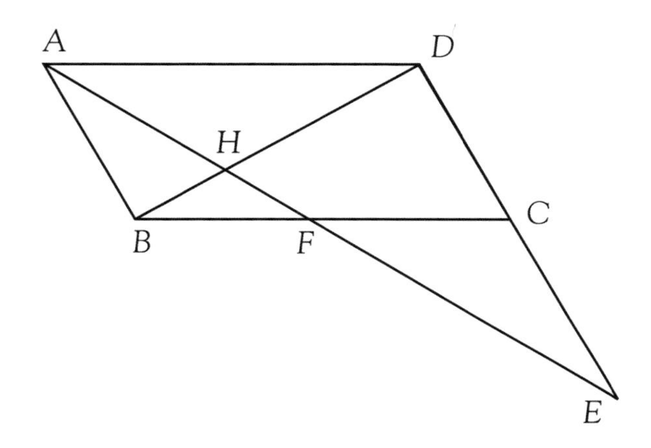

Another AA Proof

Given: $ABCD$ is a parallelogram.

Prove: $\frac{BH}{HA} = \frac{DH}{HE}$

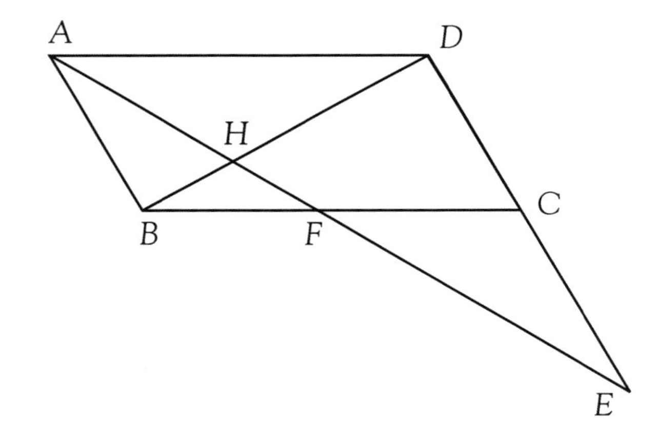

PAP Similarity

There is another shortcut theorem to similarity. If you can orient the triangles in such a way that two pairs of corresponding sides have a constant ratio and the angles between those sides are equal in measure, then the triangles are similar. This theorem is known as PAP similarity, standing for proportion-angle-proportion.

Given: $\frac{PD}{DO} = \frac{PE}{ER}$

Prove: $\overline{DE} \parallel \overline{OR}$

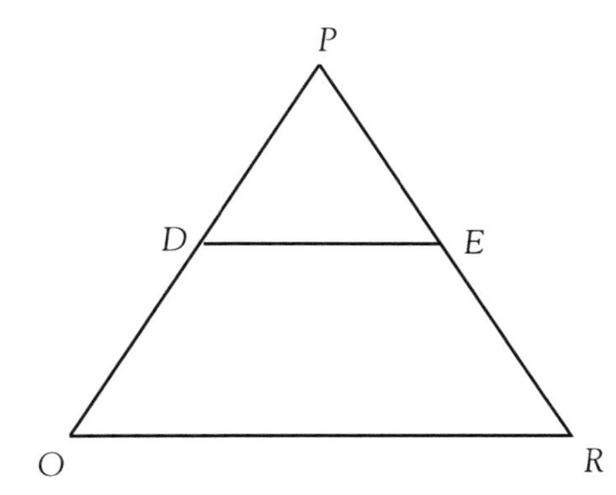

Sine Ratio

In a right triangle, the sine ratio for an acute angle is the ratio $\frac{\text{opposite}}{\text{hypotenuse}}$.

Use the triangle on the right to complete problems 1 and 2.

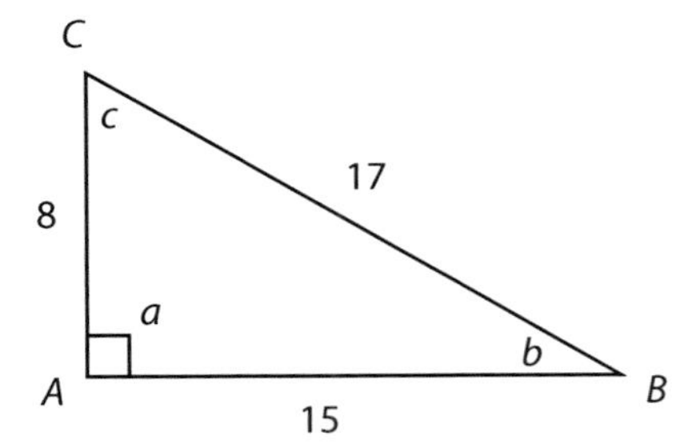

1. Find sin b.

2. Find sin c.

3. Emilio drew a 45°–45°–90° right triangle. If the sine ratio is $\frac{4}{4\sqrt{2}}$, what are the lengths of the hypotenuse and legs of the triangle?

Cosine Ratio

In a right triangle, the cosine ratio for an acute angle is the ratio $\frac{\text{adjacent}}{\text{hypotenuse}}$.

Use the triangle below to complete problems 1 and 2.

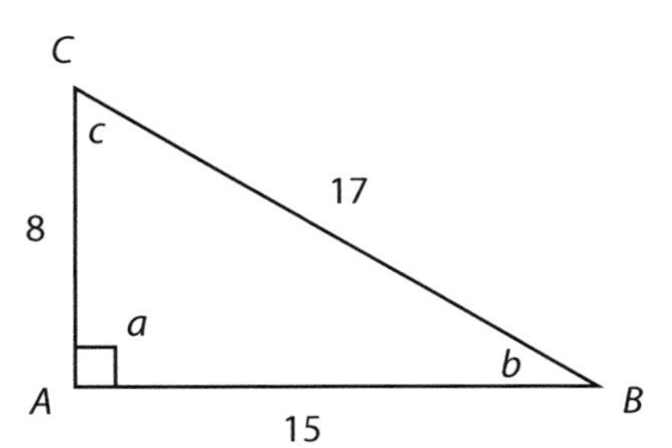

1. Find cos b.

2. Find cos c.

3. Penelope drew a 30°–60°–90° right triangle. If the cosine ratio of the 30° angle is $\frac{\sqrt{3}}{2}$, what are the lengths of the hypotenuse and the other leg of the triangle?

Tangent Ratio

In a right triangle, the tangent ratio for an acute angle is the ratio $\frac{\text{opposite}}{\text{adjacent}}$.

Use the triangle below to complete problems 1 and 2.

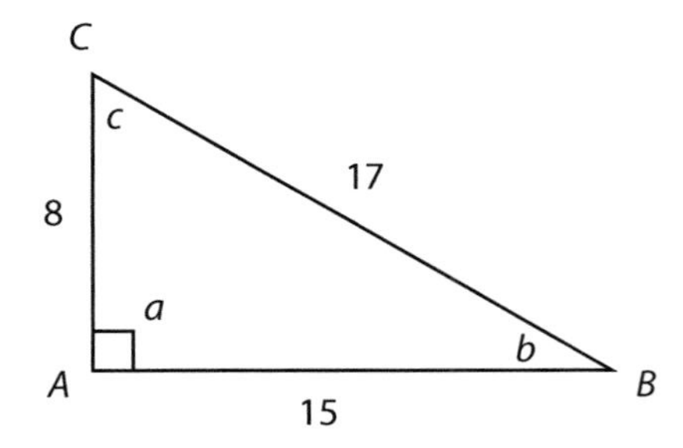

1. Find tan *b*.

2. Find tan *c*.

3. What kind of right triangle has a tangent ratio of 1 for both acute angles? Justify your answer.

Cosecant Ratio

In a right triangle, the cosecant ratio for an acute angle is the ratio $\frac{\text{hypotenuse}}{\text{opposite}}$.

Use the triangle below to complete problems 1 and 2.

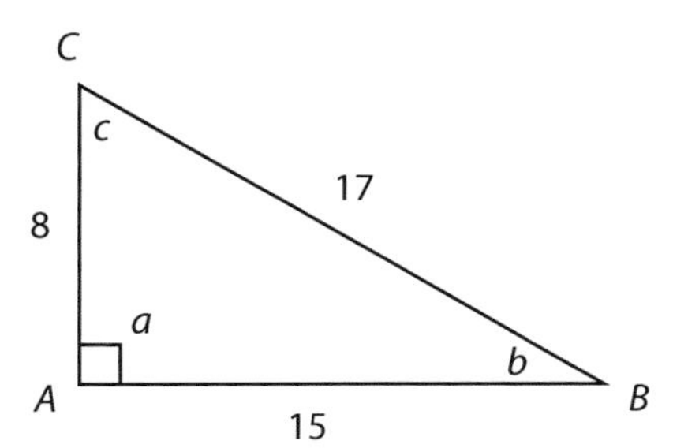

1. Find csc b.

2. Find csc c.

3. Describe the relationship between the sine ratio and the cosecant ratio.

Secant Ratio

In a right triangle, the secant ratio for an acute angle is the ratio $\frac{\text{hypotenuse}}{\text{adjacent}}$.

Use the triangle below to complete problems 1 and 2.

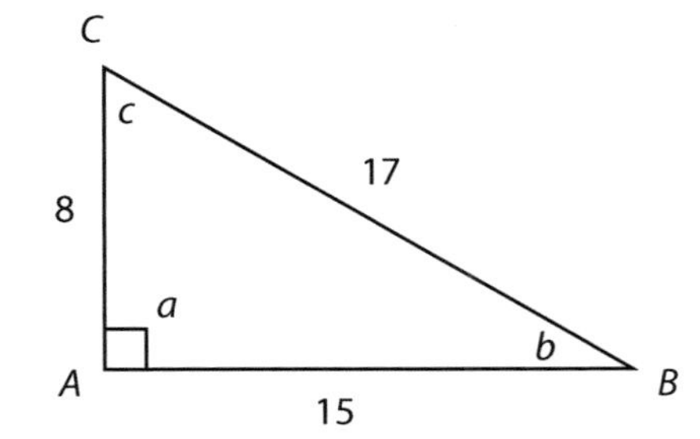

1. Find sec b.

2. Find sec c.

3. Describe the relationship between the cosine ratio and the secant ratio.

Cotangent Ratio

In a right triangle, the cotangent ratio for an acute angle is the ratio $\frac{\text{adjacent}}{\text{opposite}}$.

Use the triangle at right to complete problems 1 and 2.

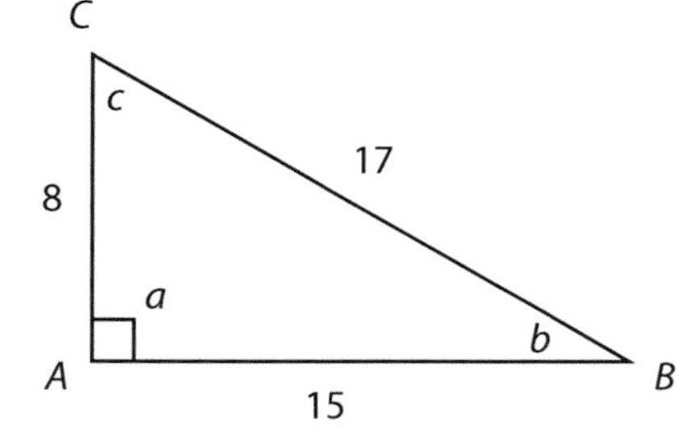

1. Find cot *b*.

2. Find cot *c*.

3. Describe the relationship between the tangent ratio and the cotangent ratio.

Law of Sines

The law of sines states that in any acute triangle, the three ratios between the sines of the angles and the lengths of the opposite sides are equal.

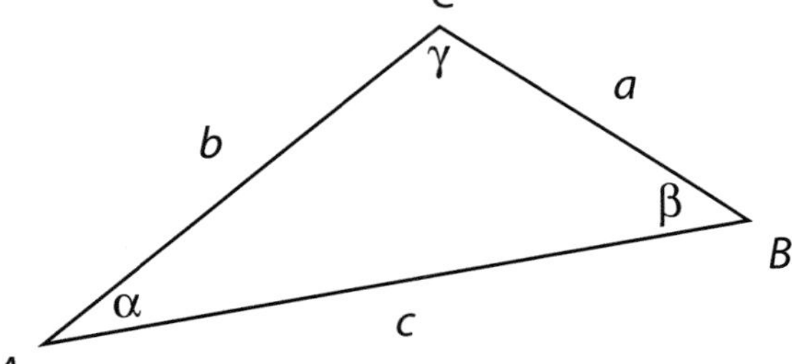

For the triangle on the right, $\dfrac{\sin\alpha}{a} = \dfrac{\sin\beta}{b} = \dfrac{\sin\gamma}{c}$.

For the triangle below, use the law of sines to find the length of $\overline{BC}$. Round your answer to the nearest hundredth.

C

84°

B

Law of Cosines

The law of cosines states that in acute triangle ABC:

$$c^2 = a^2 + b^2 - 2ab\cos\gamma$$
$$b^2 = a^2 + c^2 - 2ac\cos\beta$$
$$a^2 = b^2 + c^2 - 2bc\cos\alpha$$

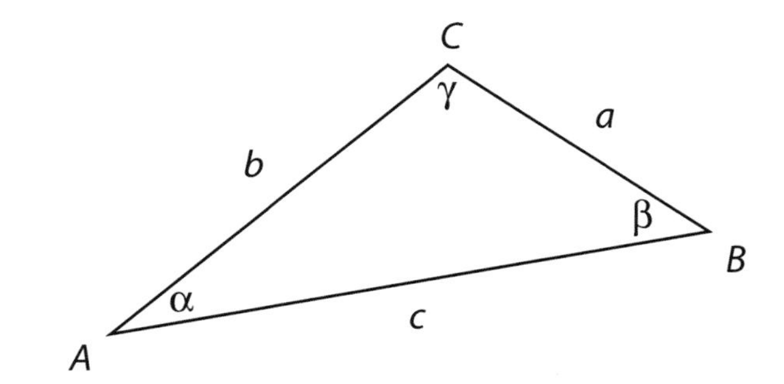

For the triangle below, find the length of $\overline{AB}$ using the law of cosines.

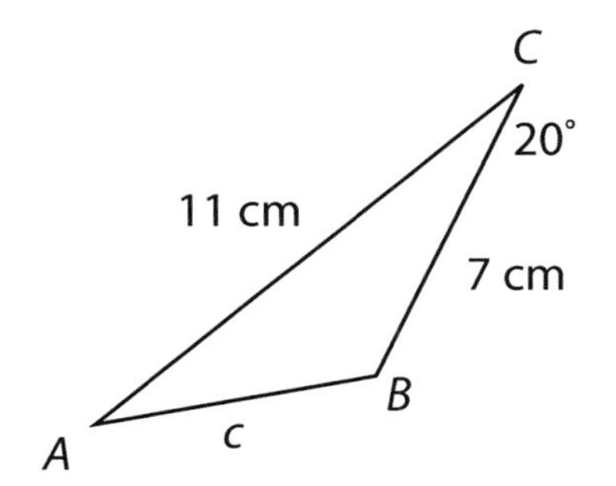

Angle of Elevation

A telephone pole casts a shadow 30 feet long on the ground. If the angle of elevation is 55°, the height of the telephone pole is 42.8 feet.

$$\tan 55° = \frac{\text{height}}{30 \text{ ft}}$$

$$\text{height} = 30 \tan 55°$$

$$\text{height} = 42.8 \text{ ft}$$

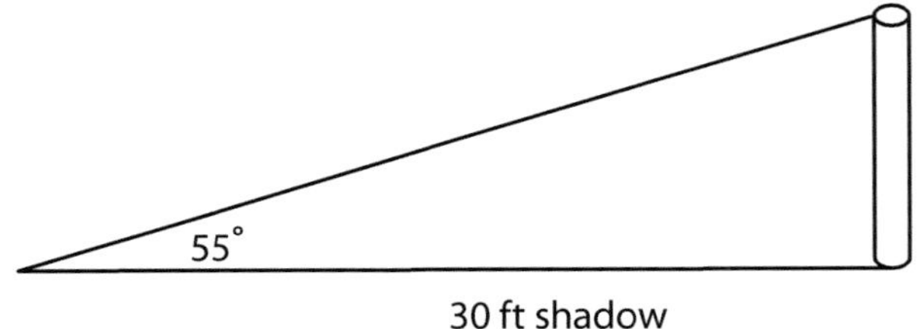

Use trigonometry ratios to find the length of the sight line from the ground to the top of the telephone pole. Show your work.

Angle of Depression

Julia rides in a hot-air balloon that is at an elevation of 2,000 feet. She spots her friend Cho on the ground. If the angle of depression between Julia and Cho is 15°, the distance of the sight line between Julia and Cho is 7,727.4 feet.

$$\sin 15^\circ = \frac{2,000}{x}$$

$$x \sin 15^\circ = 2,000$$

$$x = 7,727.4 \text{ ft}$$

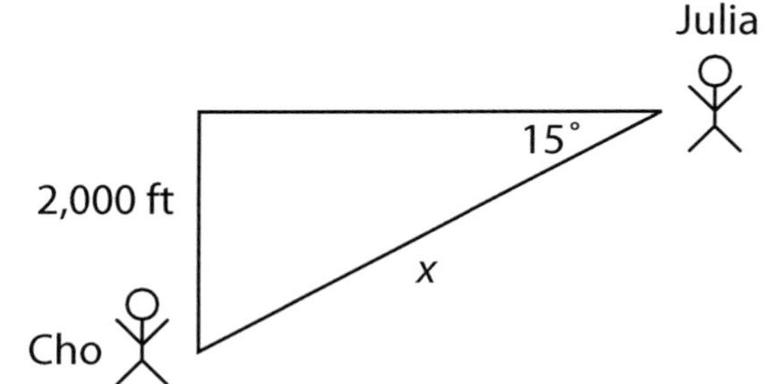

If the angle of elevation from Cho to Julia is 15°, what can you determine about the relationship between the angle of depression and the angle of elevation?

Part 3: Circles

Overview

Circles

- Understand and apply theorems about circles.

Circles and Inscribed Angles

An angle that is formed by two *chords*, such that the vertex of the angle is on the arc of the circle, is called an *inscribed angle*. Because they are formed by chords, inscribed angles also intercept the arc; however, the relationship is not one-to-one. Inscribed angles intercept an arc that measures exactly one half of the degree measure of the angle. So, a 48° inscribed angle intercepts 24° of the arc. Fill in the box below using the information given.

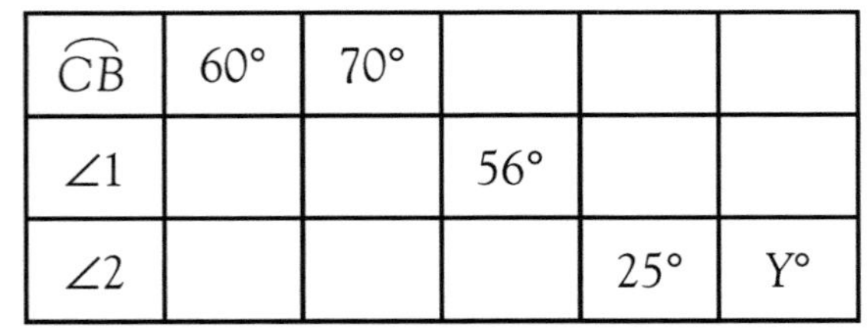

$\overset{\frown}{CB}$	60°	70°			
∠1			56°		
∠2				25°	Y°

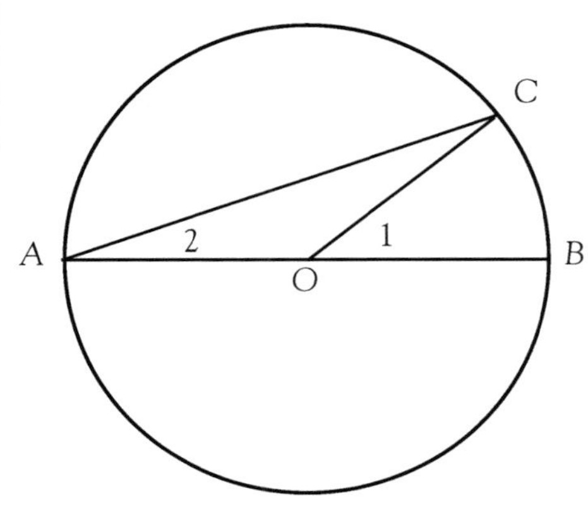

Using Inscribed Angles

In the circle below, there are two chords that intersect inside the circle, but not at the center. The angles formed at this intersection are neither central angles nor inscribed angles. However, given the measure of the two opposite pairs of arc, it is possible to determine the size of the angles at the intersection.

Given: $\overset{\frown}{AB} = 48°$

$\overset{\frown}{CD} = 96°$

Find: the measure of $\angle BED$

(*Hint:* Draw a chord that will give you inscribed angles that intercept the known arcs.)

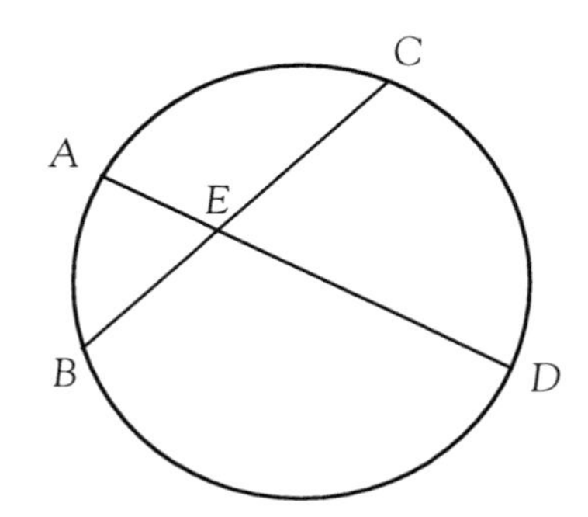

Corollary 1 to Inscribed Angles

The two ends of a chord can be said to intercept an arc as well. Corollary 1 to the inscribed angle theorem states that two chords of equal length intercept arcs of equal measure. Use what you know about inscribed angles and congruence to prove this corollary.

Given: $AB = CD$

Prove: $\widehat{AB} = \widehat{CD}$

(*Hint:* Start by drawing $\overline{AO}$, $\overline{BO}$, $\overline{CO}$, and $\overline{DO}$, and considering the triangles.)

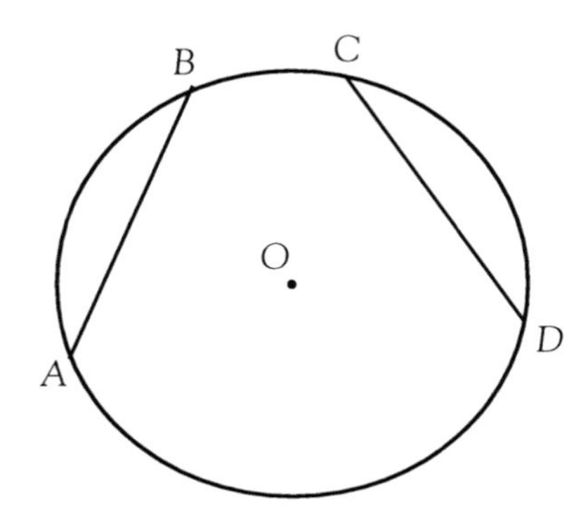

Tangent Lines to a Circle

A line that is *tangent* to a circle intersects that circle at exactly one point. A tangent line is always perpendicular to the radius that goes through its tangent point. Use a compass to draw a few circles on your paper. Draw a tangent line to one of the circles and then draw a radius of that circle from the tangent point to the center point. Try the same activity with your other circles. Use your protractor to confirm that the tangent line is always perpendicular to the radius to that tangent point.

Corollary to the Tangent Theorem

The diagram below shows a circle with two tangents drawn in that intersect at point C. Corollary 1 states that intersecting tangent segments are equal in length. By drawing $\overline{AO}$, $\overline{BO}$, and $\overline{CO}$, and using what you know about tangents being perpendicular to radii and HL congruence, prove corollary 1 about tangents.

Given: $\overline{AC}$ and $\overline{BC}$ are tangent to circle O.

Prove: $AC = BC$

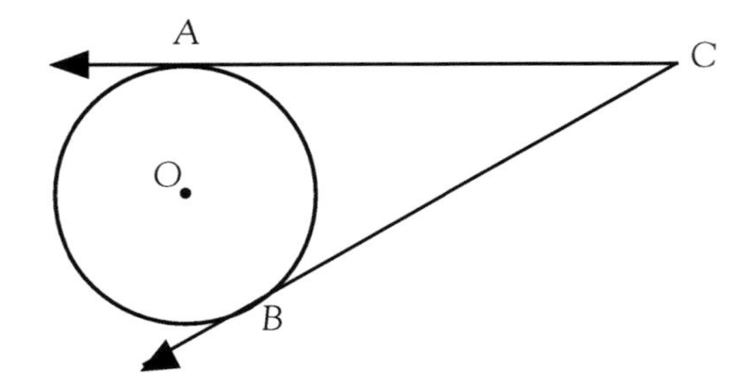

Using 30°–60°–90° Triangles in Circles

Using a protractor and a ruler, draw an equilateral triangle. Drop perpendiculars from two vertices of this triangle to the opposite sides. The point where they meet is the center of the triangle. Place the sharp point of your compass on this point and open to a radius such that the pencil point of your compass touches a vertex of the triangle. Draw a circle. Your circle should pass through all three vertices of the triangle. This triangle is said to be inscribed in the circle. By using several 30°–60°–90° calculations, we can determine that the length of the circle's radius can be calculated by multiplying the length of the triangle's side by $\frac{\sqrt{3}}{3}$. What do you suppose would be the length of the side of an equilateral triangle inscribed in a circle of radius 6?

Corollary 2 to Inscribed Angles

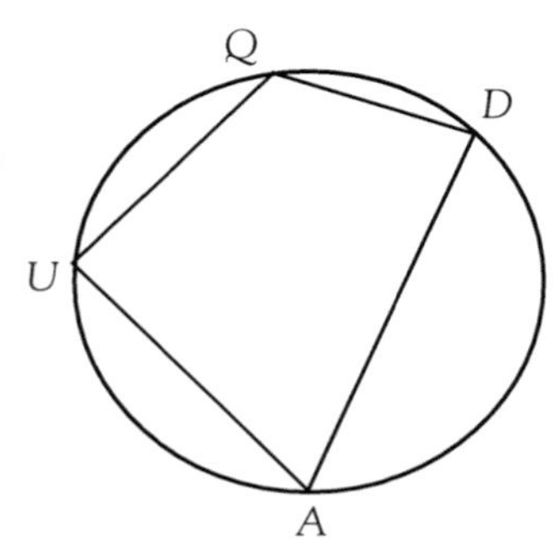

Other polygons can be inscribed in a circle. When a quadrilateral is inscribed, its opposite angles are always *supplementary*. Fill in the blanks below in the proof of this corollary.

Prove: $m\angle QUA + m\angle QDA = 180°$

Statement	Reason
1. $m\angle QUA = \frac{1}{2}(\widehat{QDA})$	Inscribed ∠'s = $\frac{1}{2}$(intercepted arc)
2. $m\angle QDA = \frac{1}{2}(\widehat{QUA})$	______________________
3. $m\angle QUA + m\angle QDA = \frac{1}{2}(\widehat{QDA} + \widehat{QUA})$	______________________
4. $\widehat{QUA} + \widehat{QDA} = 360°$	______________________
5. ______________________	Substitution of values in steps 3 and 4
6. $m\angle QUA + m\angle QDA = 180°$	______________________

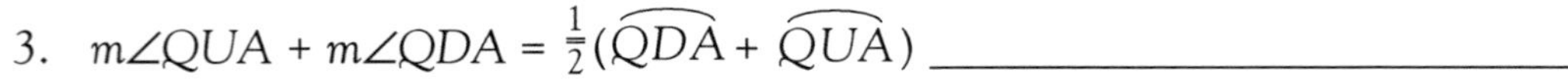

Part 4: Expressing Geometric Properties with Equations

Overview

Expressing Geometric Properties with Equations

- Use coordinates to prove simple geometric theorems algebraically.

Parallelograms

A **parallelogram** is a quadrilateral in which both pairs of opposite sides are parallel.

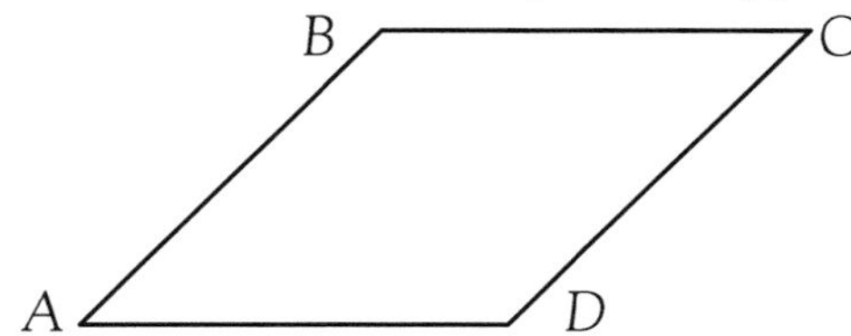

A diagonal of a parallelogram separates the parallelogram into two congruent triangles.

Plot each set of given points on the coordinate plane. Determine whether or not you can connect the points to create a parallelogram. If you can, then construct the diagonals of the parallelogram.

1. $J(1, 2)$, $K(1, 5)$, $L(4, 2)$, $M(4, 5)$

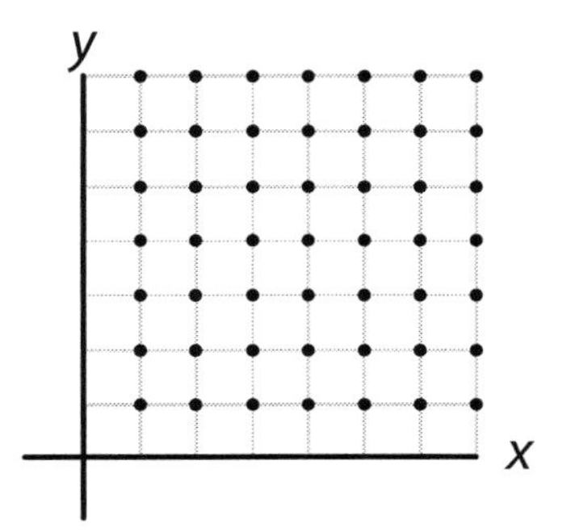

2. $A(3, 1)$, $B(7, 1)$, $C(5, 4)$

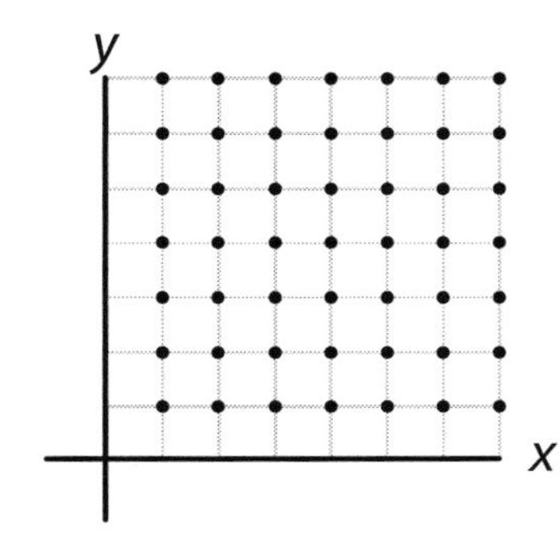

Slopes of Parallel and Perpendicular Lines

If two nonvertical lines are parallel, then their slopes are equal. If two lines (neither vertical nor horizontal) are perpendicular, then the product of their slopes is –1. In other words, the slopes of perpendicular lines are negative reciprocals of each other.

1. If the slope of line D is 4, what is the equation of the line that passes through the point (8, –1) and is perpendicular to line D?

2. If the slope of line K is $\frac{2}{3}$, what is the equation of the line that passes through the point (9, 7) and is parallel to line K?

3. The slope of line A is $\frac{5}{6}$ and the slope of line B is $\frac{-24}{20}$. Are the lines parallel, perpendicular, or neither?

Finding Area Using Given Coordinates

Using your knowledge of the distance formula, find the area of rectangle *ABCD* that has the following coordinates: $A(1, 1)$, $B(-7, -1)$, $C(-8, 3)$, and $D(0, 5)$.

Part 5: Geometric Measurement and Dimension

Overview

Geometric Measurement and Dimension

- Explain volume formulas and use them to solve problems.

Volume of a Cylinder

The volume V of a right circular cylinder with base area B and altitude of length h is given by $V = Bh$ or $V = \pi r^2 h$.

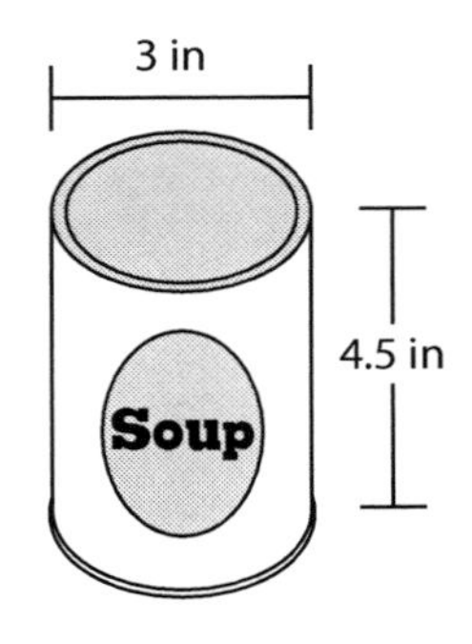

Use the can of soup shown on the right to answer the questions that follow.

1. What is the volume of the soup can? Round your answer to the nearest tenth.

2. If a second cylinder has a diameter of 4.5 inches and a height of 3 inches, does it have the same volume as the soup can? Justify your answer.

Volume of a Cone

The volume of a cone is given by $V = \frac{1}{3}\pi r^2 h$, where r is the radius of the base and h is the altitude of the cone.

Find the volume of the cone shown below. Round your answer to the nearest tenth.

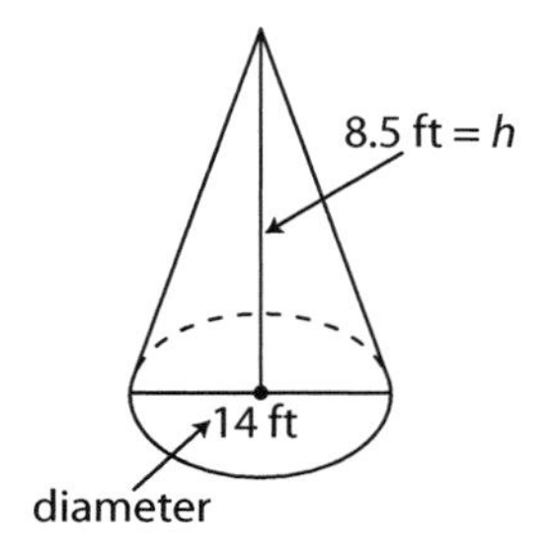

Volume of a Sphere

The volume V of a sphere with radius of length r is given by $V = \frac{4}{3}\pi r^3$. Look at the example below.

Example

Find the volume of the sphere.

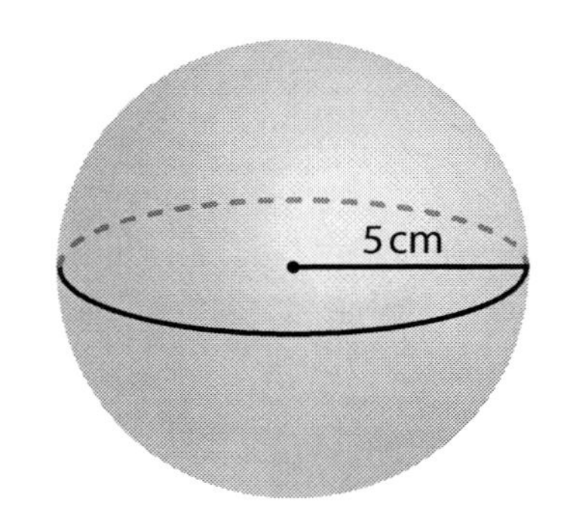

$$V = \frac{4}{3}\pi r^3$$

$$V = \frac{4}{3}\pi(5^3)$$

$$V = 166.7\pi = 523.6 \text{ cm}^3$$

Antonio has a beach ball that needs to be filled with air. If the diameter of the beach ball is 12 inches, what is the volume of the beach ball when it is filled with air?

Volume of a Pyramid

The volume V of a pyramid with a base area B and an altitude of length h is $V = \frac{1}{3}Bh$.

Find the volume of a regular hexagonal pyramid whose base edges have a length of 6 cm and whose altitude measures 15 cm.

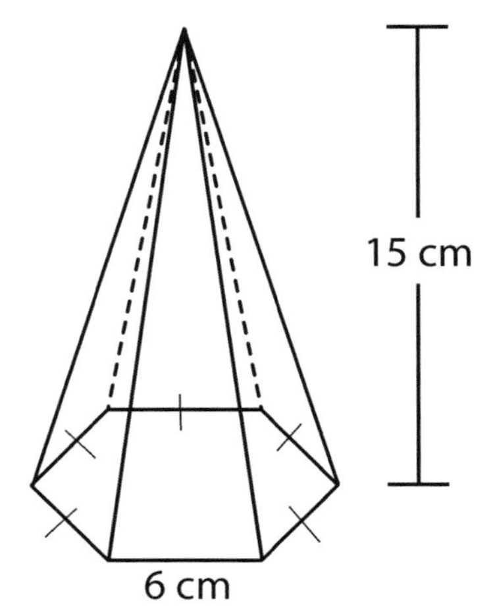

Volume of a Composite Figure

The volume of a right rectangular prism is $V = lwd$. The volume of a cylinder is found by the formula $V = \pi r^2 h$, where r is the radius of the cylinder and h is the height. The base of a trophy is made from a cylinder on top of a cube. The diameter of the cylinder is equal to the width of the cube. Find the volume of the trophy base using the information given in the diagram below.

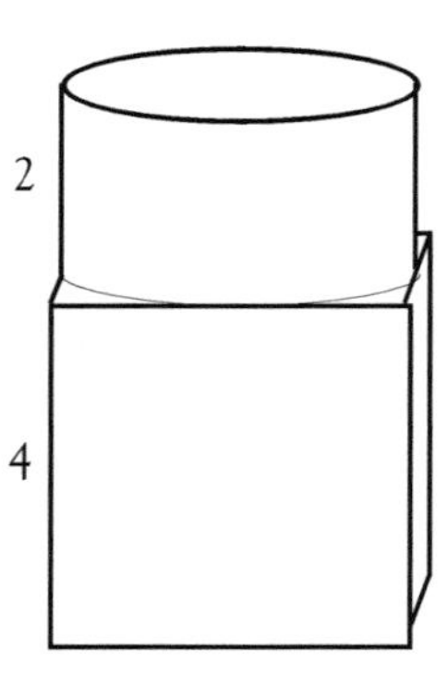

The Volume of a Regular Square Pyramid

The volume of a pyramid is given by the formula $V = \frac{1}{3}$(Area of the Base) × (Height). In this diagram, the distance from the point *E* at the top of the pyramid down to the center of the square base of the pyramid is 5 units, and the distance from *E* to the midpoint of one of the sides of the square base is 7 units. Using this information, answer the following:

a. What is the area of the square *ABCD* at the base of the pyramid?

b. What is the volume of the pyramid?

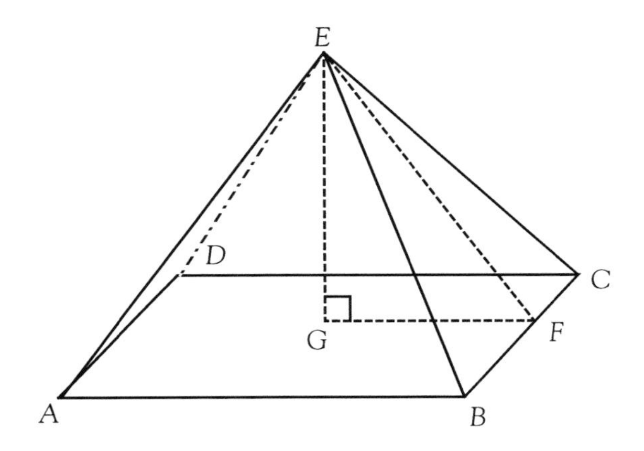

The Cone

The cone is made by drawing a circle, finding a point, P, directly above the center of the circle, and then connecting all points on the arc of the circle to that point P. The surface area of a cone is given in two parts. It is the area of the circular base plus one half the circumference of that circle, multiplied by the slant height of the cone. The volume of the cone is found using the formula $V = \frac{1}{3}\pi r^2 h$. In the diagram below, $m\angle APO = 30°$ and $AP = 8$. Determine the surface area and volume of this cone.

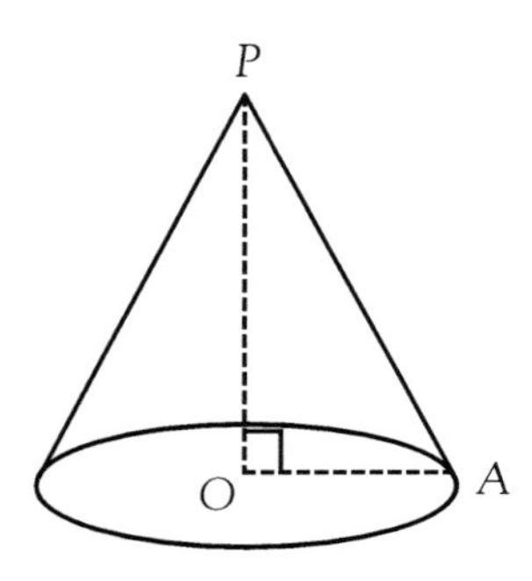

The Sphere

A sphere is a ball or the set of all points in three dimensions that are equidistant from a point called the center. The surface area of a sphere is given by the formula $SA = 4\pi r^2$. The volume of a sphere is given by the formula $V = \frac{4}{3}\pi r^3$. In the diagram of a sphere below, given that the length *PE* is 8, find the surface area and volume of the sphere.

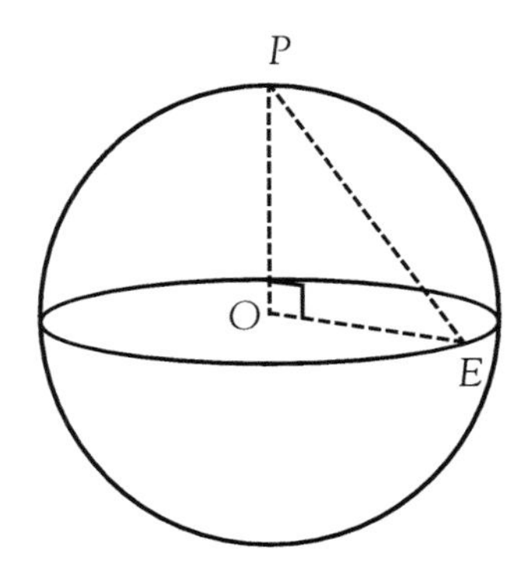

Spheres

A **sphere** is a set of points in space that are at a fixed distance from the center of the sphere. A sphere can be empty or solid.

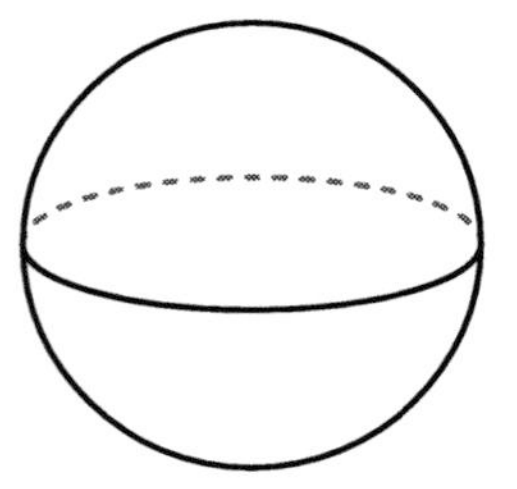

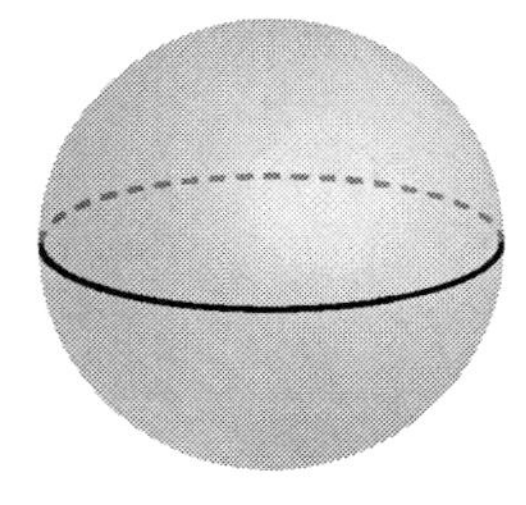

Ryhanna buys a piece of candy that is a sphere, covered in chocolate. If she cuts the candy in half, what will the cross section of the candy look like? Draw a picture for your answer.

Cross Sections of Three-Dimensional Figures

A **cross section** is the intersection of a solid and a plane. You can think of a cross section as the result of cutting a solid and viewing the inside layers of the solid.

Draw the vertical and horizontal cross sections of each solid below.

1\.

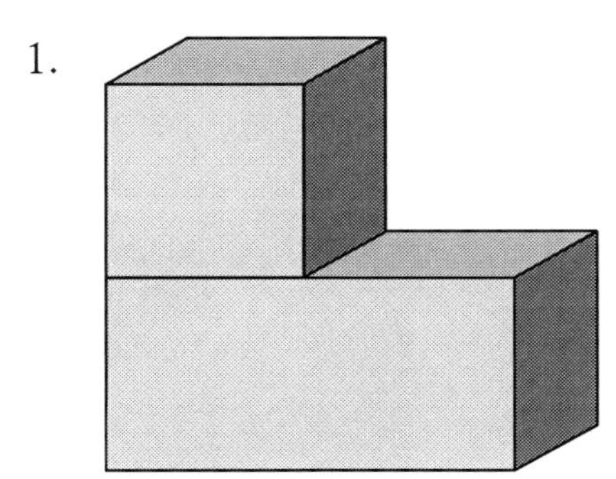

2\.

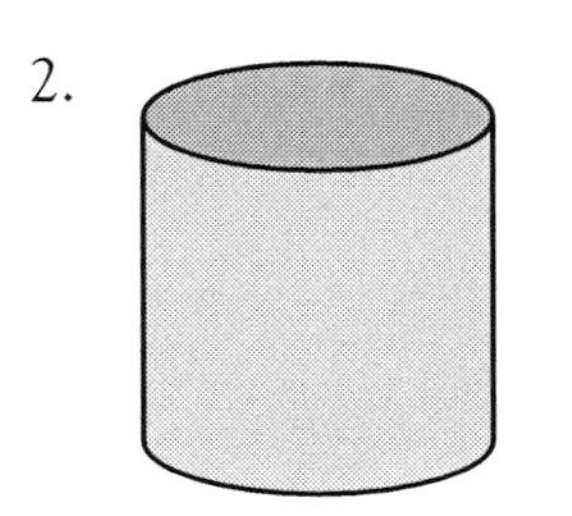

Part 1: Congruence

1.

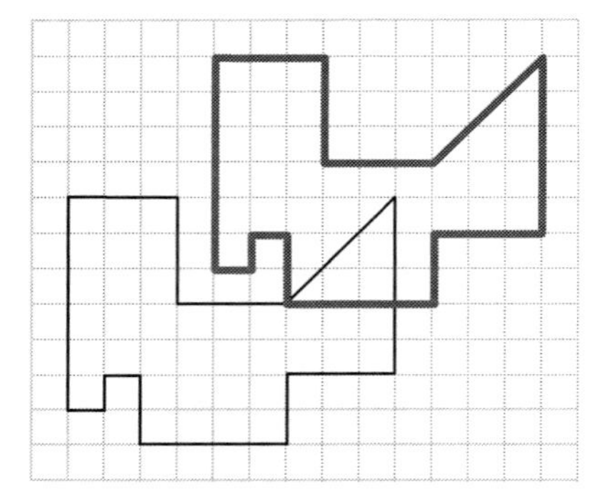

2.

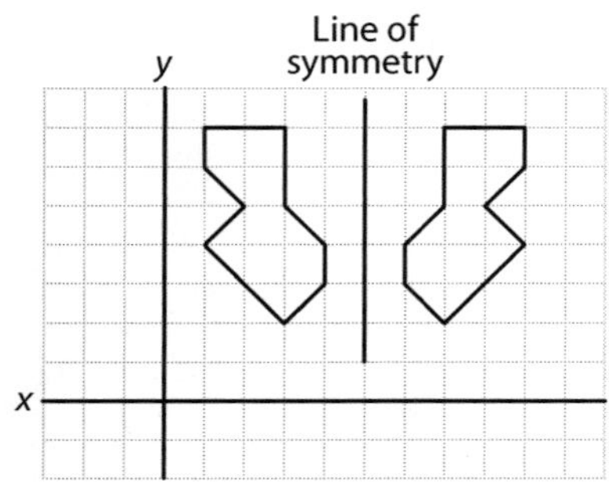

3. 1.

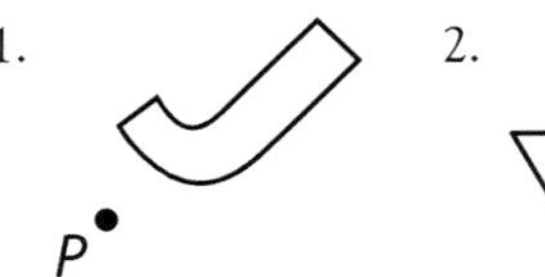

2. P

4.

Statement	Reason
1. $AB = CB$	Given
2. $\overline{BO}$ bisects $\angle ABC$.	Given
3. $m\angle ABO = m\angle CBO$	Definition of $\angle$ bisection
4. $BO = BO$	Reflexive property
5. $\Delta COB \cong \Delta AOB$	SAS
6. $CO = AO$	CPCT
7. $m\angle 1 = m\angle 2$	$2 =$ sides $\rightarrow 2 = \angle$'s

5.

Statement	Reason
1. $LO = LR$	Given
2. $RI = OI$	Given
3. $LI = LI$	Reflexive property
4. $\Delta LOI \cong \Delta LRI$	SSS
5. $m\angle LOI = m\angle LRI$	CPCT

6. Draw in the segment $\overline{AO}$ in this diagram. AO is equal in length to itself by the reflexive property. Because we are given that CO = *HO* and CA = *HA*, we can say that $\Delta CAO \cong \Delta HAO$ by the SSS congruence theorem. Because they are corresponding parts of these congruent triangles, we know that $m\angle ACO = m\angle AHO$. Also, $m\angle MOC = m\angle IOH$ because they are vertical angles and vertical angles are equal in measure. From this we can conclude that $\Delta MOC \cong \Delta IOH$. From this congruence we know that corresponding parts MO = *IO*. As these two segments are sides of the same triangle, the triangle *IOM* is isosceles and $m\angle OMI = m\angle OIM$.
7. It is given that $WI = WS$ and $IL = SL$. Also, $WL = WL$ by the reflexive property. These three side lengths allow us to determine that ΔWIL is congruent to ΔWSL. Because they are corresponding parts of congruent triangles, $m\angle IWO = m\angle SWO$. It is also clear that $WO = WO$. These facts allow the use of the SAS congruence theorem to state that $\Delta IWO \cong \Delta SWO$ and conclude that $IO = SO$ using CPCT.
8. It is given that $YN = AN$ and $OY = OA$. By the reflexive property, it is true that $ON = ON$. Using these three facts and the SSS congruence theorem, it is clear that $\Delta NOA \cong \Delta NOY$. By corresponding parts of these congruent triangles, $m\angle YNO = m\angle ANO$. Because $RN = RN$ by the reflexive property, the SAS theorem can be used to state that $\Delta RAN \cong \Delta RYN$ and we can conclude that $RY = RA$ by CPCT.
9.

Statement	Reason
1. $SH = SA$	Given
2. $HN = AN$	Given
3. $SN = SN$	Reflexive property
4. $\Delta SHN \cong \Delta SAN$	SSS
5. $m\angle HNS = m\angle ANS$	CPCT
6. $m\angle HNW = m\angle ANF$	Given
7. $m\angle WNS = m\angle FNS$	Angle addition (steps 5 and 6)
8. $m\angle HSN = m\angle FSN$	CPCT
9. $\Delta SNW \cong \Delta SNF$	ASA
10. $SW = SF$	CPCT

10. In this problem, because $m\angle MEA = 90° = m\angle AGM$, $m\angle MEA = m\angle AGM$ by substitution. It is also given that $m\angle IMA = m\angle IAM$. MA = MA by the reflexive property. These three pieces of information and the AAS theorem mean that $\Delta MGA \cong \Delta AEM$. Conclude, therefore, by CPCT that $EA = GM$.

11.

Statement	Reason
1. O is midpoint of $\overline{EN}$.	Given
2. $EO = ON$	Definition of midpoint
3. $m\angle ILE = 90°$	Given
4. $m\angle NIL = 90°$	Given
5. $m\angle ILE = m\angle NIL$	Substitution (steps 3 and 4)
6. $m\angle ION = m\angle LOE$	Vertical $\angle$'s are =.
7. $\Delta LOE \cong \Delta ION$	AAS
8. $LO = OI$	CPCT
9. $m\angle LON = m\angle EOI$	Vertical $\angle$'s are =.
10. $\Delta EOI \cong \Delta NOL$	SAS
11. $EI = LN$	CPCT

12. It is given that $IC = EC$. Also, $IE = IE$ by the reflexive property. Also given is that $m\angle ISE = 90° = m\angle ELI$, so $m\angle ISE = m\angle ELI$ by substitution. These three pieces of information with the AAS theorem allow the conclusion that $\Delta ELI \cong \Delta ISE$. By corresponding parts of congruent triangles, we conclude that $m\angle OIE = m\angle OEI$.

13. By copying a line, we may extend $\overline{AN}$ beyond N to Y, such that $NY = HT$. Draw $\overline{YE}$ to create a triangle. We know that $m\angle ENA = 90°$ and that $\overleftrightarrow{AY}$ is straight, so $m\angle ENY = 90°$. Because $HT = NY$, $m\angle ENY = m\angle ENA = 90°$ and $BH = EN$ (given), ΔBTH and ΔEYN are congruent by HS congruence. Therefore, $BT = EY$. Since BT also equals EA (given), $EY = EA$. ΔEAY is now known to be an isosceles triangle. Using the isosceles theorems, $m\angle EAY = m\angle EYA$. From the congruence statement above, $m\angle BTH = m\angle EYA$, therefore, $m\angle EAN = m\angle BTH$. ΔBTH and ΔEAN are now congruent by the AAS congruence theorem.

14.

Statement	Reason
1. $AF = GC$	Given
2. $EF = DG$	Given
3. $m\angle EAF = 90°$	Given
4. $m\angle DCG = 90°$	Given
5. $\Delta AFE \cong \Delta CGD$	HS
6. $m\angle AFE = m\angle CGD$	CPCT
7. $m\angle AFE = m\angle BFG$	Vertical $\angle$'s are =.
8. $m\angle BFG = m\angle CGD$	Substitution (steps 6 and 7)
9. $m\angle BGF = m\angle CGD$	Vertical $\angle$'s are =.
10. $m\angle BFG = m\angle BGF$	Substitution (steps 8 and 9)
11. $BF = BG$	2 = $\angle$'s → 2 = sides
12. $BE = BD$	Line addition (2, 11)
13. $m\angle FED = m\angle GDE$	2 = sides → 2 = $\angle$'s

15.

Statement	Reason
1. $AY = RF$	Given
2. $m\angle YAF = 90°$	Given
3. $m\angle FRY = 90°$	Given
4. $m\angle YAF = m\angle FRY$	Substitution (steps 2 and 3)
5. $FY = FY$	Reflexive property
6. $\Delta FAY \cong \Delta YRF$	HS
7. $m\angle AYF = m\angle RFY$	CPCT
8. $LY = LF$	2 = $\angle$'s → 2 = sides

16. It is given that $\overline{GO}$ bisects $\angle MGX$, which means that $m\angle MGO = m\angle XGO$. Also, by the reflexive property, $WG = WG$. Also given is that both $\angle WMG$ and $\angle WXG$ are right angles, so ΔWMG and ΔWXG are congruent by HS theorem. Because of corresponding parts of congruent triangles, $m\angle MWO = m\angle XWO$ and $WM = WX$. By the reflexive property, $WO = WO$, so, by SAS congruence, $\Delta WOM \cong \Delta WOX$. By corresponding parts of congruent triangles, $m\angle WOM = m\angle WOX$. Because MOX is straight, $m\angle WOM + m\angle WOX = 180°$. By substitution, we can conclude that $m\angle WOM + m\angle WOM = 180°$. Dividing this equation by 2 yields $m\angle WOM = 90°$.

17.

Statement	Reason
1. $SN = SC$	Given
2. $m\angle SIC = 90°$	Given
3. $m\angle SKN = 90°$	Given
4. $NC = NC$	Reflexive property
5. $\Delta NIC \cong \Delta CKN$	HL
6. $NI = CK$	CPCT

18. 1. yes
    2. yes
    3. no
19. 1. No. The congruent angle indicated is not between the congruent sides.
    2. No. The congruent angle indicated is not between the congruent sides.
    3. Yes. The congruent angle indicated is between the congruent sides.
    4. Yes. The congruent angle indicated is between the congruent sides.
20. 1. Yes. The congruent side indicated is between the congruent angles.
    2. No. The congruent side indicated is not between the congruent angles.
    3. Yes. The congruent side indicated is between the congruent angles.
    4. Yes. The congruent side indicated is between the congruent angles.
21.

Statement	Reason
1. $UE = EC$	Given
2. $m\angle CEL = 90°$	Given
3. $m\angle UEL = 90°$	Given
4. $m\angle UEL = m\angle CEL$	Substitution (steps 3 and 4)
5. $EL = EL$	Identity statement
6. $\Delta EUL \cong \Delta ECL$	SAS

22. We are given that GE = EO. We also know by the identity property that *EM* is equal to itself. We are also given that $m\angle GEM = m\angle OEM$. Because these three pieces of information represent two sides and the included angle, respectively, of two triangles, the triangles are congruent by the side-angle-side theorem. Because the triangles are congruent, we can conclude that the other corresponding parts of the triangles are also equal in measure. Therefore, GM = OM.
23.

Statement	Reason
1. O is midpoint of $\overline{AB}$.	Given
2. AO = BO	Definition of midpoint
3. O is midpoint of $\overline{CD}$.	Given
4. CO = DO	Definition of midpoint
5. $m\angle AOC = m\angle BOD$	Vertical $\angle$'s are =.
6. $\Delta AOC \cong \Delta BOD$	SAS
7. AC = BD	Corresponding sides are $\cong$.

24. Triangle pairs *a* and *e* are congruent by ASA.

25.

Statement	Reason
1. $m\angle WAN = m\angle YAN$	Given
2. $m\angle WNA = 90°$	Given
3. $m\angle YNA = 90°$	Given
4. $m\angle WNA = m\angle YNA$	Substitution (steps 2 and 3)
5. $AN = AN$	Reflexive property
6. $\Delta WAN \cong \Delta YAN$	ASA
7. $WA = YA$	Corresponding Parts of Congruent Triangles

26.

Statement	Reason
1. $m\angle LOR = m\angle IOR$	Given
2. $m\angle LRO = m\angle IRO$	Given
3. $RO = RO$	Reflexive property
4. $\Delta LOR \cong \Delta IOR$	ASA
5. $LO = IO$	CPCT

27.

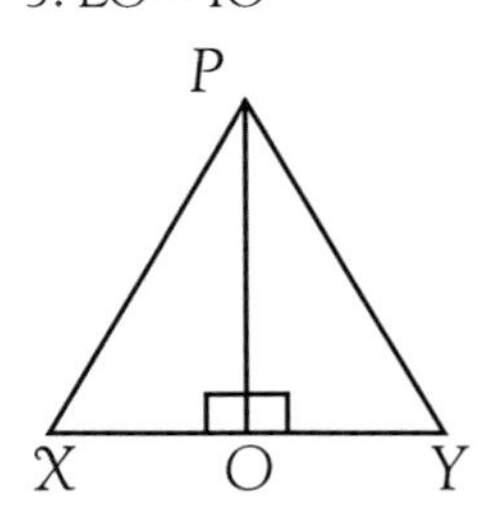

Statement	Reason
1. O is midpoint of $\overline{XY}$.	Given
2. $XO = YO$	Definition of midpoint
3. $\overline{PO} \perp \overline{XY}$	Given
4. $m\angle POX = 90°$	Definition of perpendicular
5. $m\angle POY = 90°$	Definition of perpendicular
6. $m\angle POY = m\angle POX$	Substitution (steps 4 and 5)
7. $PO = PO$	Reflexive property
8. $\Delta POY \cong \Delta POX$	SAS

28.

Statement	Reason
1. $m\angle GAM = 90°$	Given
2. $m\angle NAM = 90°$	Given
3. $m\angle NAM = m\angle GAM$	Substitution (steps 1 and 2)
4. $MA = MA$	Reflexive property
5. $GA = GN$	Given
6. $\Delta MAG \cong \Delta MAN$	SAS
7. $MG = MN$	CPCT
8. $m\angle EMG = m\angle NMC$	Given
9. $m\angle EGM = m\angle CNM$	Given
10. $\Delta MEG \cong \Delta MCN$	ASA
11. $ME = MC$	CPCT

29.

Statement	Reason
1. $BH = HY$	Given
2. $EH = HN$	Given
3. $m\angle BHE = m\angle YHN$	Vertical ∠'s are =.
4. $\Delta BEH \cong \Delta YNH$	SAS
5. $BE = NY$	CPCT
6. $m\angle EBT = m\angle AYN$	Given
7. $\Delta BET \cong \Delta YNA$	ASA
8. $BT = AY$	CPCT

30. The triangles will always be congruent. In fact, knowing that three sides of one triangle are of equal length, respectively, to three sides of another triangle, guarantees that the triangles are congruent. A proof lies in reconstructing one triangle along one of the sides of the other triangle, constructing a line between vertices and then using what you know about isosceles triangles to show that SSS information leads to an SAS congruence.

31.

Statement	Reason
1. $m\angle 1 = m\angle 4$	Given
2. $m\angle 2 = m\angle 5$	Given
3. $m\angle 1 + m\angle 2 + m\angle BCA = 180°$	$\sum \angle$'s $\Delta = 180°$
4. $m\angle 4 + m\angle 5 + m\angle DFE = 180°$	$\sum \angle$'s $\Delta = 180°$
5. $m\angle 1 + m\angle 2 + m\angle BCA = m\angle 4 + m\angle 5 + m\angle DFE$	Substitution (steps 3 and 4)
6. $m\angle 2 + m\angle BCA = m\angle 5 + m\angle DFE$	Subtraction (steps 5 and 1)
7. $m\angle BCA = m\angle DFE$	Subtraction (steps 6 and 2)
8. $AB = DE$	Given
9. $\Delta ABC \cong \Delta DEF$	ASA

32.
1. $m\angle 2 = 70°$
2. $m\angle 3 = 110°$

33.

Statement	Reason
1. $m\angle 1 + m\angle 2 = 180°$	Definition of a straight line
2. $m\angle 2 + m\angle 3 = 180°$	Definition of a straight line
3. $m\angle 1 + m\angle 2 = m\angle 2 + m\angle 3$	Transitive property
4. $m\angle 2 = m\angle 2$	Reflexive property
5. $m\angle 1 = m\angle 3$	Subtraction (steps 3 and 4)

34. Check students' drawings.

35.

Statement	Reason
1. $\overline{AI} \parallel \overline{EP}$	Given
2. $m\angle ITC + m\angle TCP = 180°$	EV
3. $m\angle ITC + m\angle ATC = 180°$	Definition of a straight line
4. $m\angle ATC + m\angle ITC = m\angle ITC + m\angle TCP$	Substitution (steps 2 and 3)
5. $m\angle ITC = m\angle ITC$	Reflexive property
6. $m\angle ATC = m\angle TCP$	Subtraction (steps 4 and 5)
7. $m\angle ATC + m\angle ECT = 180°$	EV
8. $m\angle ATC + m\angle ECT = m\angle ATC + m\angle ITC$	Substitution (steps 3 and 7)
9. $m\angle ATC = m\angle ATC$	Reflexive property
10. $m\angle ECT = m\angle ITC$	Subtraction (steps 8 and 9)

36.

Statement	Reason
1. $m\angle ECT = m\angle ITC$	Given
2. $m\angle ITC + m\angle ATC = 180°$	Definition of a straight line
3. $m\angle ECT + m\angle ATC = 180°$	Substitution (steps 1 and 2)
4. $\overline{AI} \parallel \overline{EP}$	EV

37.

Statement	Reason
1. $\overline{AI} \parallel \overline{EP}$	Given
2. $m\angle ATC = m\angle PCT$	$\parallel \rightarrow$ Alt. Int. $\angle$'s =
3. $m\angle ATC = m\angle KTI$	Vertical $\angle$'s are =.
4. $m\angle KTI = m\angle PCT$	Substitution (steps 2 and 3)
5. $m\angle PCT = m\angle MCE$	Vertical $\angle$'s are =.
6. $m\angle KTI = m\angle MCE$	Substitution (steps 4 and 5)

38.

Statement	Reason
1. $m\angle KTI = m\angle MCE$	Given
2. $m\angle KTI = m\angle ATC$	Vertical $\angle$'s are =.
3. $m\angle ATC = m\angle MCE$	Substitution (steps 1 and 2)
4. $m\angle MCE = m\angle PCT$	Vertical $\angle$'s are =.
5. $m\angle ATC = m\angle PCT$	Substitution (steps 3 and 4)
6. $\overleftrightarrow{AI} \parallel \overleftrightarrow{EP}$	Alt. Int. $\angle$'s = $\rightarrow \parallel$

39.

Statement	Reason
1. $\overleftrightarrow{AI} \parallel \overleftrightarrow{EP}$	Given
2. $m\angle ATC = m\angle PCT$	$\parallel \rightarrow$ Alt. Int. $\angle$'s =
3. $m\angle ATC = m\angle KTI$	Vertical $\angle$'s are =.
4. $m\angle KTI = m\angle PCT$	Substitution (steps 2 and 3)

40.

Statement	Reason
1. $m\angle KTI = m\angle PCT$	Given
2. $m\angle ATC = m\angle KTI$	Vertical $\angle$'s are =.
3. $m\angle ATC = m\angle PCT$	Substitution (steps 1 and 2)
4. $\overleftrightarrow{AI} \parallel \overleftrightarrow{EP}$	Alt. Int. $\angle$'s = $\rightarrow \parallel$

41. We are given that E is the midpoint of $\overline{DF}$, which means that $ED = EF$. Also given is that $\overline{OE} \perp \overline{DF}$, meaning that angles DEO and FEO are right angles, which allows us to substitute and then show that $m\angle DEO = m\angle FEO$. By the reflexive property, $EO = EO$. By these three facts we can conclude, using the SAS congruence theorem, that ΔDEO is congruent to ΔFEO. Because they are corresponding parts of congruent triangles, we now know that $\angle ODE$ is equal in measure to $\angle OFE$. Because $\overleftrightarrow{AC} \parallel \overleftrightarrow{DF}$, $m\angle AOD = m\angle ODE$ because they are alternate interior angles. By substitution then, $m\angle AOD = m\angle OFE$. $\angle OFE$ is also an alternate interior angle to $\angle COF$. By substitution then, $m\angle COF = m\angle AOD$. However, $\angle COF$ is a vertical angle to $\angle BOA$, meaning that they are equal in measure, so we can substitute again and conclude that $m\angle BOA = m\angle DOA$. If these two angles are equal in measure and added together form $\angle BOD$, we can conclude that $\overline{AO}$, forming the division of $\angle BOD$, is the angle bisector of $\angle BOD$.

42.

Statement	Reason
1. $\overline{AT}$ bisects $\angle KAE$.	Given
2. $\angle KAT = \angle EAT$	Definition of $\angle$ bisection
3. $\overline{AT} \parallel \overline{IE}$	Given
4. $m\angle TAE = m\angle IEA$	$\parallel \rightarrow$ Alt. Int. $\angle$'s =
5. $m\angle KAT = m\angle IEA$	Substitution (steps 2 and 4)
6. $m\angle IAE + m\angle EAT + m\angle TAK = 180°$	Definition of straight line
7. $m\angle IAE + m\angle AIE + m\angle IEA = 180°$	$\sum \angle$'s $\Delta = 180°$
8. $m\angle IAE + m\angle EAT + m\angle TAK = m\angle IAE + m\angle AIE + m\angle IEA$	Substitution (steps 6 and 7)
9. $m\angle IAE + m\angle TAK = m\angle IAE + m\angle AIE$	Subtraction (steps 8 and 4)
10. $m\angle IAE = m\angle IAE$	Reflexive property
11. $m\angle TAK = m\angle AIE$	Subtraction (steps 9 and 10)
12. $m\angle EAT = m\angle AIE$	Substitution (steps 2 and 11)
13. $m\angle IEA = m\angle AIE$	Substitution (steps 4 and 12)
14. $AE = AI$	2 = $\angle$'s $\rightarrow$ 2 = Sides

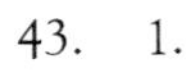

43. 1.

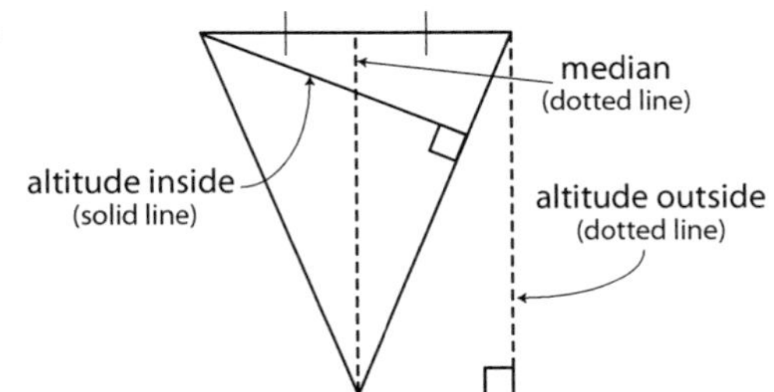

2.

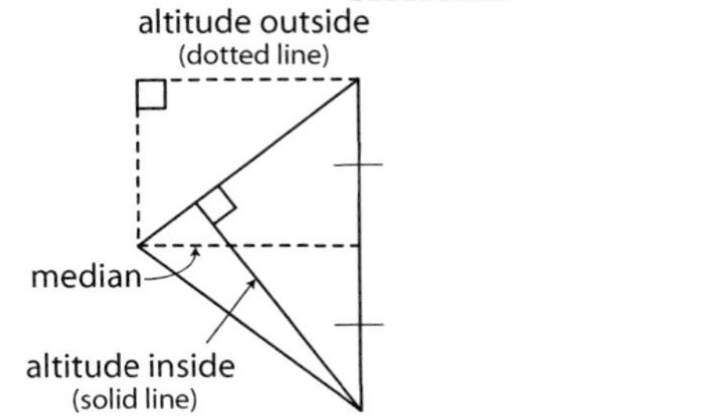

3.

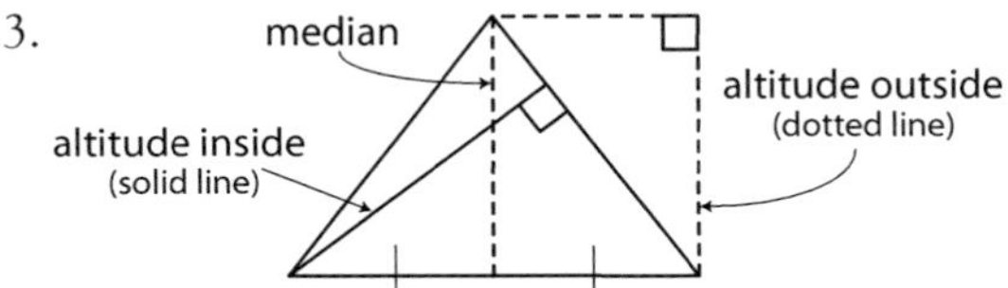

44. If all three medians are concurrent to a point, then the point is the centroid. Look for marks and arcs in the students' drawings to make sure that they did all three constructions.

45. The angles form a 180° arc.

46. Because the line at the base of the triangle is straight, $m\angle 3 + m\angle 4 = 180°$. Inside the triangle, we know that $m\angle 1 + m\angle 2 + m\angle 3 = 180°$. Because they both measure 180°, we can substitute and know that $m\angle 1 + m\angle 2 + m\angle 3 = m\angle 3 + m\angle 4$. Subtracting $m\angle 3$ from each side, we conclude that $m\angle 1 + m\angle 2 = m\angle 4$.

47.

Statement	Reason
1. Name remaining $\angle$'s in triangles $\angle 5$ (top) and $\angle 6$ (bottom).	Construction
2. $m\angle 1 + m\angle 3 + m\angle 5 = 180°$	$\sum \angle$'s $\Delta = 180°$
3. $m\angle 2 + m\angle 4 + m\angle 6 = 180°$	$\sum \angle$'s $\Delta = 180°$
4. $m\angle 1 + m\angle 3 + m\angle 5 = m\angle 2 + m\angle 4 + m\angle 6$	Substitution (steps 2 and 3)
5. $m\angle 1 = m\angle 2$	Given

6. $m\angle 3 + m\angle 5 = m\angle 4 + m\angle 6$ — Subtraction (steps 4 and 5)
7. $m\angle 5 = m\angle 6$ — Vertical ∠'s are =.
8. $m\angle 3 = m\angle 4$ — Subtraction (steps 6 and 7)

48. i. Because $AB = AC$, and $m\angle BAC = m\angle BAC$, we can use SAS to state that $\Delta BAC \cong \Delta CAB$. Because they are corresponding parts of congruent triangles, $m\angle ABC = m\angle ACB$.

ii.

Statement	Reason
1. $m\angle ABC = m\angle ACB$	Given
2. $BC = BC$	Reflexive property
3. $\Delta BAC \cong \Delta CAB$	ASA
4. $AB = AC$	CPCT

49.
1. F; A parallelogram can also be a square or a rhombus.
2. T; All rectangles fit the definition of a parallelogram.
3. F; A rhombus is a parallelogram that does not have a right angle.

50.
1. a. T; A square always fits the definition of a rectangle.
   b. F; A rectangle does not always fit the definition of a square.
2. The parallelogram is a rectangle because not all four sides are congruent. Instead, two pairs of sides are congruent: x and x, and $x + 2$ and $x + 2$.

51.
1. a rhombus
2. a square or a rectangle

52.

Statement	Reason
1. $m\angle CEF = m\angle BFE$	Given
2. $EF = EF$	Reflexive property
3. $\overleftrightarrow{AB} \parallel \overleftrightarrow{CD}$	Given
4. $m\angle CFE = m\angle BEF$	‖ → Alt. Int. ∠'s =
5. $\Delta CEF \cong \Delta BFE$	ASA
6. $BF = CE$	CPCT

53.

Statement	Reason
1. $\overleftrightarrow{PA} \parallel \overleftrightarrow{LR}$	Given
2. $m\angle APR = m\angle LRA$	‖ → Alt. Int. ∠'s =
3. $PA = LR$	Given
4. $PR = PR$	Reflexive property
5. $\Delta CEF \cong \Delta BFE$	ASA
6. $BF = CE$	CPCT

54.

Statement	Reason
1. Draw $\overline{AE}$.	Construction
2. $\overline{AL} = \overline{AL}$	Reflexive property
3. $\overline{AL} \parallel \overline{XE}$	Given
4. $m\angle LAE = m\angle XEA$	$\parallel \rightarrow$ Alt. Int. $\angle$'s =
5. $\overline{AX} \parallel \overline{LE}$	Given
6. $m\angle XAE = m\angle LEA$	$\parallel \rightarrow$ Alt. Int. $\angle$'s =
7. $\Delta XAE \cong \Delta LEA$	ASA = ASA $\angle$ Δ's $\cong$
8. $m\angle 1 = m\angle 2$	CPCT

55.

Statement	Reason
1. Draw $\overline{FN}$.	Construction
2. $\overline{FN} = \overline{FN}$	Reflexive property
3. $\overline{FR} \parallel \overline{AN}$	Given
4. $m\angle RFN = m\angle ANF$	$\parallel \rightarrow$ Alt. Int. $\angle$'s =
5. $\overline{FA} \parallel \overline{RN}$	Given
6. $m\angle AFN = m\angle RNF$	$\parallel \rightarrow$ Alt. Int. $\angle$'s =
7. $\Delta AFN \cong \Delta RNF$	ASA
8. $FR = AN$	CPCT

56.

Statement	Reason
1. $\overline{ZA} \parallel \overline{CH}$	Given
2. $m\angle AZO = m\angle CHO$	$\parallel \rightarrow$ Alt. Int. $\angle$'s =
3. $\overline{ZC} \parallel \overline{AH}$	Given
4. $m\angle CZO = m\angle AHO$	$\parallel \rightarrow$ Alt. Int. $\angle$'s =
5. $ZH = ZH$	Reflexive property
6. $\Delta CZH \cong \Delta AHZ$	ASA
7. $AH = ZC$	CPCT
8. $m\angle ZCO = m\angle HAO$	$\parallel \rightarrow$ Alt. Int. $\angle$'s =
9. $\Delta ZCO \cong \Delta HAO$	ASA
10. $ZO = OH$	CPCT

57. 1. a. $\overrightarrow{JK}$; 15/16 cm

b. $\overleftarrow{RS}$; 2/16 cm

c. $\overleftrightarrow{QV}$; 0.75 cm

2. a. 12-cm line with ends labeled G and *H*

b. 5.5-cm line with ends labeled A and *B*

c. 16.2-cm line with ends labeled *L* and O

58. Check constructions.

59. Check constructions.

60. Check constructions.

61. Check constructions.

62. Check constructions.

63.

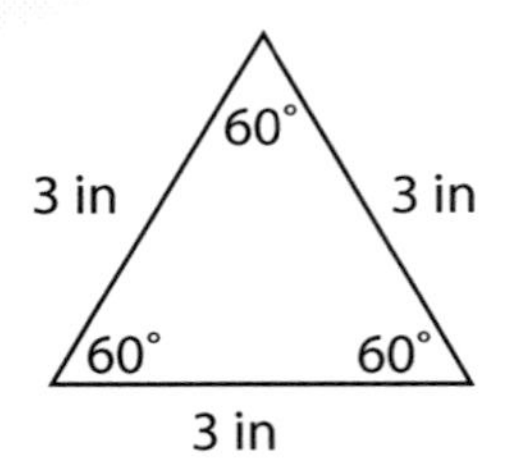

Part 2: Similarity, Right Triangles, and Trigonometry

64. Check students' constructions.

65. 1. Yes. They have the same shape.
    2. No. They do not have the same shape.
    3. Yes. They have the same shape.
    4. No. They do not have the same shape. One is a regular hexagon; the other is an irregular hexagon.

66. $AB = 150$ feet; therefore, the stage is 150 feet across.

67. $\Delta JKL \sim \Delta NPM$, $m\angle NMP = 38.21°$

68. $AN = 4\frac{1}{3}$ and $EM = 7\frac{2}{3}$

69.

Statement	Reason
1. $BC \parallel DE$	Given
2. $m\angle ABC = m\angle ADE$	$\parallel \rightarrow$ Corresponding $\angle$'s =
3. $\angle BAC = \angle DAE$	Reflexive property
4. $\Delta ABC \sim \Delta ADE$	AA = AA $\rightarrow \Delta$'s ~

70. $BD = 27\frac{\sqrt{5}}{5}$

71. We are given that $\overline{AB} \parallel \overline{DE}$. Using the alternate interior angle theorem, we know $m\angle ABC = m\angle EDC$. Using the vertical angle theorem, we know that $m\angle BCA = m\angle DCE$. These two angle pairs with equal measure let us use AA to conclude $\Delta ABC \sim \Delta EDC$.

72.

Statement	Reason
1. $\overline{AC} \perp \overline{BD}$	Given
2. $m\angle 1 = m\angle 2$	Given
3. $m\angle ACD = 90°$	Given
4. $m\angle ACB = 90°$	Given
5. $\Delta BAC \sim \Delta EDC$	AA

73. Given that $ABCD$ is a parallelogram, we know a number of things about the diagram. Because $\overline{BC} \parallel \overline{AD}$, $m\angle ADB = m\angle CBD$. Because $\overline{AB} \parallel \overline{DC}$, $m\angle ABD = m\angle CDB$. These two pairs of angles that are equal in measure allow us to conclude that $\Delta ABD \sim \Delta CDB$.

74.

Statement	Reason
1. $\overline{RS} \parallel \overline{VT}$	Given
2. $m\angle RSV = m\angle TVS$	$\parallel \rightarrow$ Alt. Int. $\angle$' s =
3. $m\angle SRT = m\angle VTR$	$\parallel \rightarrow$ Alt. Int. $\angle$' s
4. $\Delta ROS \sim \Delta TOV$	AA

75. Given that $\overline{AB}$ is parallel to $\overline{DE}$, we can use the alternate interior angle theorem to show that $m\angle BAH = m\angle DEH$. Given that AD is parallel to BC, the same theorem indicates that $m\angle ABF = m\angle EAD$. From these pairs of angles, it can be seen that $\Delta ABF \sim \Delta EDA$. Using the full meaning of the triangles being similar, we can conclude that corresponding sides are in constant ratio or, in other words, $\frac{AB}{ED} = \frac{BF}{DA}$.

76. Because they are vertical angles, $m\angle AHB = m\angle EHD$. Because $\overline{AB} \parallel \overline{DE}$, $m\angle ABH = m\angle EDH$. These two pairs of equal angles indicate that the AA similarity theorem can be used: $\Delta HAB \sim \Delta HED$. Because they are corresponding sides of similar triangles, $\frac{HA}{HE} = \frac{BH}{DH}$. Using the property of proportions, we conclude that $\frac{BH}{HA} = \frac{DH}{HE}$.

77.

Statement	Reason
1. $\frac{PD}{DO} = \frac{PE}{ER}$	Given
2. $\frac{DO}{PD} = \frac{ER}{PE}$	Proportion rules
3. $\frac{(DO + PD)}{PD} = \frac{(ER + PE)}{PE}$	Proportion rules
4. $\frac{PO}{PD} = \frac{PR}{PE}$	Definition of line addition
5. $m\angle DPE = m\angle DPE$	Reflexive property
6. $\Delta PDE \sim \Delta POA$	PAP
7. $m\angle PDE = m\angle POR$	Corresponding Parts of Similar Triangles
8. $m\angle PDE + m\angle ODE = 180°$	Definition of straight line

9. $m\angle POR + m\angle ODE =$ 180° — Substitution (steps 6 and 7)

10. $\overline{DE} \parallel \overline{OR}$ — EV

78. 1. 8/17
    2. 15/17
    3. Hypotenuse = $4\sqrt{2}$; each leg of the triangle is 4.

79. 1. 15/17
    2. 8/17
    3. hypotenuse = 2; leg = 1

80. 1. 8/15
    2. 15/8
    3. A 45°–45°–90° triangle has a tangent ratio of 1 for both acute angles. This is true because both legs of the triangle have equal lengths.

81. 1. 17/8
    2. 17/15
    3. The sine ratio is the inverse of the cosecant ratio.

82. 1. 17/15
    2. 17/8
    3. The cosine ratio is the inverse of the secant ratio.

83. 1. 15/8
    2. 8/15
    3. The tangent ratio is the inverse of the cotangent ratio.

84. $BC \approx 7.87$

85. $AB \approx 5.03$

86. $\cos 55° = 30/x$

    $x \cos 55° = 30$

    $x = 52.3$ ft

87. The angle of elevation is congruent to the angle of depression.

Part 3: Circles

88.

$\overset{\frown}{CB}$	60°	70°	56°	50°	2Y°
∠1	60°	70°	56°	50°	2Y°
∠2	30°	35°	28°	25°	Y°

89. $m\angle BED = 108°$

90.

Statement	Reason
1. Draw $\overline{AO}$, $\overline{BO}$, $\overline{CO}$, and $\overline{DO}$.	Construction
2. $AO = CO$	Definition of radius
3. $BO = DO$	Definition of radius
4. $AB = CD$	Given
5. $\Delta AOB \cong \Delta COD$	SSS
6. $m\angle AOB = m\angle COD$	CPCT
7. $m\angle AOB = \widehat{AB}$	Central angle = Int. arc
8. $m\angle COD = \widehat{CD}$	Central angle = Int. arc
9. $\widehat{AB} = m\angle COD$	Substitution (steps 6 and 7)
10. $\widehat{AB} = \widehat{CD}$	Substitution (steps 8 and 9)

91. Check drawings.

92. Draw $\overline{AO}$, $\overline{BO}$, and $\overline{CO}$ in the diagram. CO is equal to itself by the reflexive property. $m\angle OAC = 90°$ and $m\angle OBC = 90°$ by the tangent theorem. $AO = BO$ because they are both radii of the same circle. By these facts we can use the HL congruence theorem to conclude that $\Delta AOC \cong \Delta BOC$, and therefore that corresponding parts AC and BC are equal in length.

93. Side = $3\sqrt{3}$

94.

Statement	Reason
1. $m\angle QUA = \frac{1}{2}(\widehat{QDA})$	Inscribed ∠'s = $\frac{1}{2}$ (intercepted arc)
2. $m\angle QDA = \frac{1}{2}(\widehat{QUA})$	Inscribed ∠'s = $\frac{1}{2}$ (intercepted arc)
3. $m\angle QUA + m\angle QDA = \frac{1}{2}(\widehat{QDA} + \widehat{QUA})$	Substitution (steps 1 and 2)
4. $\widehat{QUA} + \widehat{QDA} = 360°$	Definition of a full rotation
5. $m\angle QUA + m\angle QDA = \frac{1}{2}(360°)$	Substitution (steps 3 and 4)
6. $m\angle QUA + m\angle QDA = 180°$	Multiplication of fractions

Part 4: Expressing Geometric Properties with Equations

95. 1. Yes, it is a parallelogram.

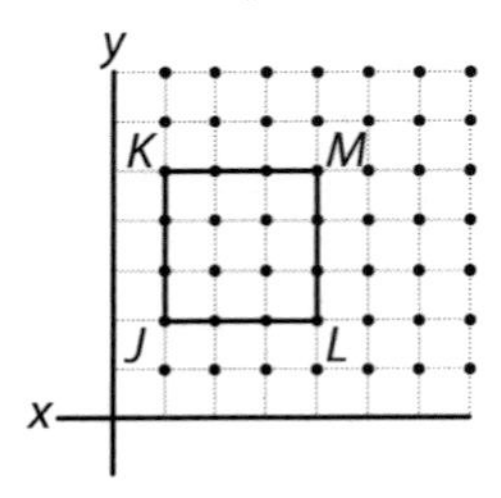

2. No, it is not a parallelogram because only three points are given.

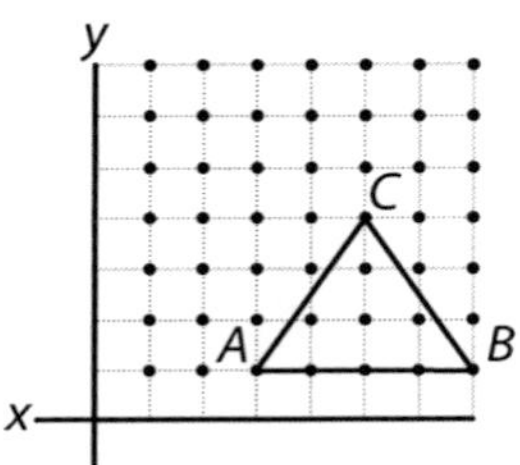

96. 1. $y = -1/4x + 1$
    2. $y = 2/3x + 1$
    3. perpendicular
97. 34 square units

Part 5: Geometric Measurement and Dimension

98. 1. $V = 31.8 \text{ in}^3$
    2. No, they have different volumes. The volume of the second cylinder is $V = 47.7 \text{ in}^3$.
99. $V = 436.2 \text{ ft}^3$
100. $V = 288\pi = 904.8 \text{ in}^3$
101. $V = 270\sqrt{3} \text{ cm}^3$
102. Volume is $8\pi + 64$ cubic units.
103. a. 96 square units
     b. 160 cubic units
104. Surface area of the cone is 48π square units.
     Volume of the cone is $\pi 64\frac{\sqrt{3}}{3}$ cubic units.
105. Surface area of the sphere is 128π square units.
     Volume of the sphere is $(\pi 512\sqrt{2})/3$ cubic units

Answer Key

106.

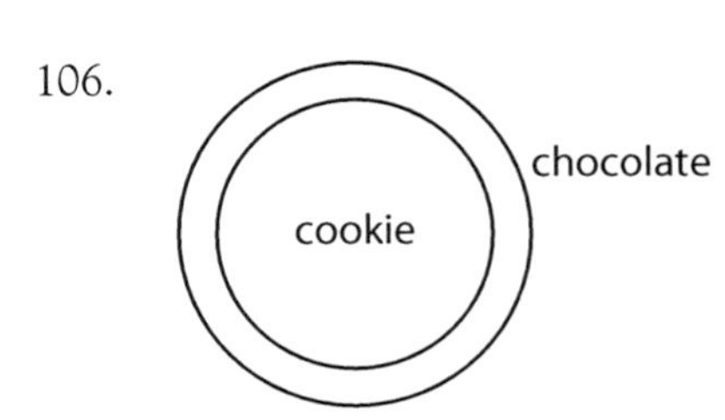

107. 1. horizontal cross section:

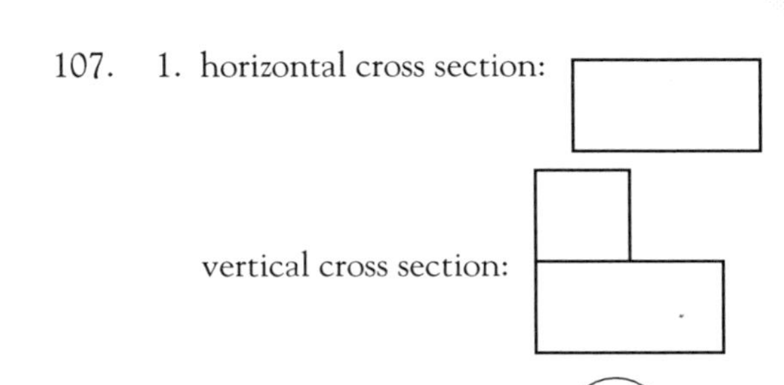

vertical cross section:

2. horizontal cross section:

vertical cross section: